Population and Climate Change

Population and Climate Change provides the first systematic in-depth treatment of links between two major themes of the 21st century: population growth and associated demographic trends such as aging, and climate change.

The book examines the role of demographic factors in greenhouse gas emissions and asks how population affects the ability of societies and institutions to respond to the potential impacts of climate change. Based on this review, it considers whether climate change strengthens the case for population policies. The book provides overview chapters aimed at nonspecialists on climate change, population, and population–economy–environment interactions that provide sufficient context for understanding the interdisciplinary analysis in the second half of the book. It is written by a multidisciplinary team of authors from the International Institute for Applied Systems Analysis, who integrate both natural science and social science perspectives in a way that is readable by members of both communities.

This book will be of primary interest to researchers in the fields of climate change, demography, and economics. It will also be useful to policymakers and NGOs dealing with issues of population dynamics and climate change, and to teachers and students on courses such as environmental studies, demography, climatology, economics, earth systems science, and international relations.

Brian C. O'Neill is an assistant professor (research) at the Watson Institute for International Studies and Center for Environmental Studies, Brown University.

F. Landis MacKellar is leader of the Social Security Reform Project at the International Institute for Applied Systems Analysis (IIASA).

Wolfgang Lutz is leader of the Population Project at the International Institute for Applied Systems Analysis (IIASA) and secretary general of the International Union for the Scientific Study of Population.

The International Institute for Applied Systems Analysis

is an interdisciplinary, nongovernmental research institution founded in 1972 by leading scientific organizations in 12 countries. Situated near Vienna, in the center of Europe, IIASA has been for more than two decades producing valuable scientific research on economic, technological, and environmental issues.

IIASA was one of the first international institutes to systematically study global issues of environment, technology, and development. IIASA's Governing Council states that the Institute's goal is: *to conduct international and interdisciplinary scientific studies to provide timely and relevant information and options, addressing critical issues of global environmental, economic, and social change, for the benefit of the public, the scientific community, and national and international institutions.* Research is organized around three central themes:

– Environment and Natural Resources;
– Energy and Technology;
– Population and Society.

The Institute now has national member organizations in the following countries:

Austria
The Austrian Academy of Sciences

Bulgaria*
The Bulgarian Committee for IIASA

Finland
The Finnish Committee for IIASA

Germany**
The Association for the Advancement
of IIASA

Hungary
The Hungarian Committee for Applied
Systems Analysis

Japan
The Japan Committee for IIASA

Kazakhstan*
The Ministry of Science –
The Academy of Sciences

Netherlands
The Netherlands Organization for
Scientific Research (NWO)

Norway
The Research Council of Norway

Poland
The Polish Academy of Sciences

Russian Federation
The Russian Academy of Sciences

Slovak Republic*
The Slovak Committee for IIASA

Sweden
The Swedish Council for Planning and
Coordination of Research (FRN)

Ukraine*
The Ukrainian Academy of Sciences

United States of America
The US Committee for IIASA

*Associate member
**Affiliate

Population and Climate Change

Brian C. O'Neill
F. Landis MacKellar
Wolfgang Lutz

with contributions by

Lee Wexler
Astri Suhrke
Anthony McMichael
Anne Goujon
Johannes Stripple

I I A S A

PUBLISHED BY THE PRESS SYNDICATE OF THE UNIVERSITY OF CAMBRIDGE
The Pitt Building, Trumpington Street, Cambridge, United Kingdom

CAMBRIDGE UNIVERSITY PRESS
The Edinburgh Building, Cambridge CB2 2RU, UK
40 West 20th Street, New York, NY 10011-4211, USA
10 Stamford Road, Oakleigh, VIC 3166, Australia
Ruiz de Alarcón 13, 28014 Madrid, Spain
Dock House, The Waterfront, Cape Town 8001, South Africa

http://www.cambridge.org

First published 2001

Printed in the United States of America

Typeface Times 11/13.6 *System* LaTeX [Leslie Lamport]

A catalog record for this book is available from the British Library

Library of Congress Cataloging in Publication data available

ISBN 0 521 66242 7 hardback

Contents

Foreword

For most of its 25-year history, the International Institute for Applied Systems Analysis (IIASA) has conducted important original research in climate change and in demography. It is therefore particularly appropriate that scientists from this institute undertook the multidisciplinary research project that has resulted in this book.

We can expect that climate change, both natural changes and those resulting from a variety of societal activities, will lead to changes in demography. In turn, demographic changes could influence climate. *Population and Climate Change* is the first book to delve deeply into the intricate relations between shifting climates and population changes such as where people live, where they move to, how they propagate, and how they die. It is also the first book to take a dispassionate, scientific look at how population affects climate change and society's ability to adapt to it.

History is replete with examples of climate change influencing population. Twenty thousand years ago, the amount of water stored in continental glaciers reached its maximum. As a result, the sea level dropped more than 100 meters, exposing parts of the ocean floor as dry land. Shallow ocean straits became land bridges; one such formation, the Bering Strait, served as a migratory route allowing Asiatic nomads to cross over to North America. This early example illustrates how natural climate change influences migration.

The impacts of climate change were particularly noticeable during medieval times. Lack of sanitation and modern technologies left people vulnerable to even small changes in the prevailing climate. In some sense, the vulnerability of people in these early times provides guidance as to the vulnerability of people living in today's developing world. H. H. Lamb, considered by many the dean of climate historians, described the wet period during the 14th and 15th centuries as a time of diseases. Diseases affected humans, animals, and crops. In England, the average expected lifetime decreased from approximately 48 years in the late 13th century to around 38 years by 1376–1400. One of the most dreadful diseases of that period was ergotism, also known as Saint Anthony's Fire, produced by ergot blight which blackened the kernels of rye in damp harvests during periods of excess rainfall. Small amounts of the black grains, when baked in bread, caused the disease. Whole villages suffered convulsions, hallucinations, gangrene of the extremities of the body, and death.

As if ergotism were not enough, the Black Death arrived in Europe around 1350, recurring periodically throughout the next few centuries. Although the origins of the Black Death are complicated, it is thought to have originated in China or Central Asia in a region where the bubonic plague is endemic, either during or immediately after exceptional rains and flooding around 1332. These floods took 13 million lives in China and the death of many million more followed as the Black Death spread across Asia and Europe over the ensuing centuries.

The Little Ice Age of the 17th and 18th centuries had a major impact on migration as agriculture failed in various parts of Europe. Scotland and Ireland, where agricultural conditions were marginal at best, suffered greatly. People from both countries moved elsewhere, many to North America. The failure of harvests in Central and Northern Europe as a result of persistent cold weather in summer forced large numbers of the population to migrate, again with many moving to North America.

Today, technological optimists make the comfortable assumption that with the advance of technology and with preventive structures in place, humankind is no longer vulnerable to shifts in climate. While this may be partly true in the advanced economies of Europe and North America, it is certainly not true for much of the developing world. But even the developed world could face problems.

The Gulf Stream, which warms much of Western Europe, could collapse if salty, cold, and dense waters fail to plunge to ocean depths in the Labrador Sea and in the seas east of Greenland and Iceland. In one scenario, a more active hydrologic cycle reduces the density of the salt water by flooding the ocean surface with fresh water from rain and melting ice. A collapse of the Gulf Stream would have profound effects on the agriculture of Western Europe from the Nordic countries south to France and Spain. The climate of Western Europe would become more continental in character, with Vienna's weather becoming more like that of present-day Kiev.

In examining the interaction between climate and population, we need to take into account the world's changing demography. As discussed in this book, the world population continues to grow, but at the same time it is also growing older. Population aging is partly the result of decreases in mortality as health conditions and medical technology advance. However, in the longer term, a massive decline in fertility will result in an ever growing graying population.

While few countries have a declining population, some 60 countries with more than 40% of the world population have fertility rates below the replacement rate (around 2.1 births per woman), and all the highly industrialized countries of Europe, North America, and Japan have fertility rates below the replacement rate.

Fertility rates have hit rock bottom in Eastern Europe and the former Soviet Union, and are declining rapidly in much of East Asia and Latin America.

The changing age structure of populations has led to regional migrations. In the United States, both aging and economic factors have led to an increase in population in the hurricane-prone regions of the Atlantic and Gulf coasts. For example, Florida's population has increased fivefold since 1950 and 80% of Florida's population now lives within 35 kilometers of the coast.

In regions of rapid economic development there have been major migrations to urban centers, which are particularly vulnerable to natural hazards because of their dependence on complex infrastructures. For example, the great ice storm of 1997 came within two power lines of depriving Montreal of all electricity, perhaps for weeks or months. Older urban centers are especially vulnerable to weather-related catastrophes. In 1992, a winter storm, not even of hurricane strength, led to an almost complete shutdown of the New York metropolitan transportation system. Much of the subway system lies below sea level and the storm surge of only 2 1/2 meters led to major flooding of the subway system. A hurricane of magnitude 4 would be accompanied by a storm surge of about 10 meters, which would flood lower Manhattan, pound the World Trade Center, and flood John F. Kennedy International Airport with approximately 3 meters of water.

The extreme scenarios I have described have a low probability, but they could happen. An urban dweller in a high-rise apartment may feel safe from weather-related hazards. But while his or her physical safety is assured, the supporting infrastructure remains much more vulnerable. Meanwhile, countries such as Honduras and Nicaragua have been economically paralyzed for years by Hurricane Mitch and its associated storm surge.

It remains uncertain whether catastrophic weather events will become more frequent and more intense in a world with a changing climate. Climate change can certainly threaten agriculture and health, as illustrated by the shutdown of the Gulf Stream scenario. Climate change is a long-term phenomenon, measured in decades and centuries. During those decades and centuries there are many opportunities for institutions to change and for the countries now most vulnerable to enter into the ranks of the developed world. But for this to come about, the interactions between climate and people, and between demographics and climatology, must be better understood. This book does a superb job of outlining the challenges that lie ahead.

Gordon MacDonald
Director
International Institute for
Applied Systems Analysis

Preface

Demographic trends and climate change will be two important themes of the 21st century. Both areas have been researched intensively, yet relatively little attention has been paid to the relationship between them. This book aims to describe that relationship.

The book is divided into two parts. In Part I, we discuss the climate outlook, demographic prospects, and economic perspectives on population, development, and the environment. These chapters take as starting points existing reviews of each field: the latest assessment of the Intergovernmental Panel on Climate Change (IPCC; Bruce *et al.*, 1996; Houghton *et al.*, 1996; Watson *et al.*, 1996); the documentation for the global population projections of the International Institute for Applied Systems Analysis (IIASA; Lutz, 1996); and a chapter on economics, demography, and the environment from an assessment of the social scientific aspects of climate change (Rayner and Malone, 1998). In Part II we analyze three major links between population and climate change: the role of population growth and structure in greenhouse gas (GHG) emissions, the effect of population growth and structure on the resilience of societies to the expected impacts of global warming, and the implications of global warming for population-related policies.

Chapter 1 reviews the natural science aspects of climate change and describes how the Earth's natural greenhouse effect is being enhanced by emissions of carbon dioxide and other GHGs originating from human activities – principally the burning of fossil fuels such as coal, oil, and natural gas. In the absence of restrictive policies, continued population growth and increases in economic output are expected to drive emission rates higher in the future. How much and how fast climate will change as a result are uncertain. The regional impacts of changes are also uncertain, but they are expected to be wide-ranging and potentially severe. Extreme climate-related events, such as floods, windstorms, and droughts, may become more intense and occur more frequently. We also trace the development of global climate change as a political issue and describe the 1997 Kyoto Protocol to the Framework Convention on Climate Change (FCCC), which, if ratified, would require emission reductions by industrialized countries.

One source of uncertainty regarding the future path of GHG emissions, and therefore eventual climate change, lies in future population trends. In Chapter 2, we review the dramatic changes that occurred in fertility, mortality, and migration over the second half of the 20th century and summarize current developments. We present a number of projections of global population trends over the coming century, identifying three near certainties: continued population growth, continued population aging, and a further tilting of the balance of population toward the currently developing countries. The key uncertainty in future global population growth is the course of fertility in the developing regions, and we review theories of fertility decline upon which projections are based. The key uncertainty in population age structure is also fertility.

Part I concludes with a summary of the state of research into population–development–environment interactions. In Chapter 3 we describe a basic neoclassical economic growth model and examine the role of population, paying particular attention to possible relationships between population growth and income. We then add the environment link, emphasizing the trade-offs between fertility, consumption of material goods, and quality of the environment. We discuss the effects of population growth on natural resources within this framework and contrast the neoclassical view with alternative points of view such as that of ecological economics. Finally, we examine a "vicious-circle model," which extends the basic population–development–environment model by incorporating poverty and the status of women. Such models have become influential in formulating policy. They describe a self-reinforcing series of responses to stress in low-income settings that can lead to a destructive spiral of poverty, high fertility, and environmental degradation. Nonetheless, such models hold out the possibility for virtuous responses as well. In a closing section, Chapter 3 discusses the economic consequences of population aging. While uncertain, these consequences are more likely to be negative than positive, and their impact on the global economy may be significant.

With Part I as a background, in Part II we examine links between population and climate change. Chapter 4 analyzes the relationship between demographic trends and GHG emissions. Most analyses of this relationship have been based on decomposing trends in demographic impact identities such as the Impact–Population–Affluence–Technology (I=PAT) model. We argue that the results of decompositions of this sort are difficult to interpret and nearly impossible to compare. However, simple linear models such as I=PAT do provide a logical starting point for sensitivity analysis, that is, comparison of baseline scenarios with alternative scenarios in which one or more of the parameters or assumptions built into the baseline has been altered. We present results of such an analysis, concluding that GHG emissions are sensitive to population growth in the long run, although not in the short run. This holds true even when possible relationships between population and affluence are accounted for or different demographic units of account, such as households, are employed. A review of population in a number of major integrated climate change assessment models supports this conclusion.

In Chapter 5 we ask whether slowing population growth might improve the ability of less developed countries (LDCs) to adapt to the expected impacts of climate change. We review research on agriculture, health, and environmental security, looking in each case at the current situation, the expected impacts of climate change, and what is known about how demographic pressures may contribute to either vicious or virtuous responses to such impacts. Our examination of agriculture includes an assessment of how population growth affects some of the major components of agricultural systems: land (including consequences of deforestation), soil, water, and technology. The section on health gives special attention to the spread of vector-borne diseases, as well as potential increases in the number and frequency of heat waves. The review of environmental security issues focuses on the concept of "environmental refugees" and whether climate change might dramatically worsen migration driven by environmental degradation. We conclude that, on balance, reduced demographic pressure in the form of lower fertility and slower population growth would make developing countries more resilient to the expected impacts of climate change in these three areas. In no area, however, could population-related policies be called core strategies – there are more direct and likely more effective measures that could be taken.

The final chapter asks whether climate change strengthens the case for population policies, especially those directly or indirectly related to fertility. We begin by outlining the justifications for climate policies and the justifications for population policies. Some of these justifications are grounded in efficiency and others in equity; some are derived from utilitarian individualism and others from ecological prudence; some are political while others are economic; some are deterministic while others are based on uncertainty. The Cairo Programme of Action (the statement that emerged from the 1994 International Conference on Population and Development) justifies population policies primarily on the basis of empowerment of women and the securing of reproductive health rights. Impacts on fertility and consequences for the environment are secondary concerns.

However, in LDCs, policies such as voluntary family planning programs and investments in girls' education are not only desirable in their own right, but also accelerate fertility decline, which may have significant benefits in the context of climate change. Slower population growth in LDCs will not only tend to reduce GHG emissions but is also likely, based on our review of the evidence, to increase the ability of societies to adapt to the impacts of climate change. Thus, while there are more direct means of mitigating emissions and strengthening institutions, population policies can be considered "win–win" strategies. Since GHG emissions and climate change represent classic externalities to individual fertility decisions, they may justify population policies that not only enable couples to have the number of children they desire, but also create conditions that may lead to lower desired family size where fertility is currently high. Justifications for investments in education of girls and the empowerment of women, whose fertility desires are often lower than men's, are therefore strengthened.

The situation in MDCs is more complicated. In countries (such as the USA) where actual fertility continues to exceed desired fertility by a significant margin, policies to reduce the number of unwanted births are also win–win in the context of climate change. Given that emissions per capita are much higher in MDCs than in LDCs (and that this gap is projected to persist), resulting reductions in GHG emissions could be significant. However, most MDCs are concerned with negative impacts of low fertility and population aging, such as stresses on pension and health care systems and unfavorable changes in labor force size and structure. These effects represent an external cost to low fertility just as reduced GHG emissions represent an external benefit. Therefore, it is unlikely that climate change will lead to any interest on the part of MDC policymakers in promoting still lower levels of desired fertility. Climate change might, nonetheless, offset the attractiveness of pronatalist policies for addressing the external costs of low fertility.

Climate change gives rise to interregional spillover and externality effects in international population policy. When LDC policymakers take actions that, directly or indirectly, result in lower fertility and slower population growth, they confer an external benefit on future generations in MDCs in the form of lower GHG emissions. This strengthens the case for continued provision of North–South financial assistance in the area of population.

We close the book by drawing attention to directions for future work. Among these are emerging concerns over the relationship between demographic factors and extreme climate-related events, population and environment in an urban setting, and aging and the environment.

Authors and Contributors

Authors

Brian O'Neill, who was principally responsible for Chapters 1 and 4, is at the Watson Institute for International Studies and the Center for Environmental Studies, Brown University, Providence, RI, USA.

Landis MacKellar, who was principally responsible for Chapters 3 and 5, is leader of the Social Security Reform Project at the International Institute for Applied Systems Analysis (IIASA), Laxenburg, Austria.

Wolfgang Lutz, who was principally responsible for Chapter 2, is leader of the Population Project at IIASA and Secretary General of the International Union for the Scientific Study of Population (IUSSP).

Contributors

Anthony McMichael, on whose writings the health section of Chapter 5 is based, is Professor of Epidemiology at the London School of Hygiene and Tropical Medicine, London, UK.

Astri Suhrke, on whose writings the environmental security section of Chapter 5 is based, is a senior researcher at the Chr. Michelsen Institute, Bergen, Norway.

Anne Goujon, who contributed the section on urbanization to Chapter 2, is a researcher in the Population Project at IIASA.

Johannes Stripple, who contributed research on extreme climate events and natural catastrophes to Chapter 1, is a Ph.D. student at the University of Lund, Sweden.

Lee Wexler, who contributed to Chapter 4 and is responsible for Appendices II and III, was formerly a researcher in the Population Project at IIASA. He is currently a computer programmer in Boston, MA, USA.

Acknowledgments

This manuscript has its origins in the four-volume Battelle Pacific Northwest Laboratories study Human Choice and Climate Change (Rayner and Malone, 1998). As the convening author responsible for the chapter on population, health, and nutrition, MacKellar commissioned contributions from Lutz, McMichael, Suhrke, O'Neill, and Wexler. As the Battelle study neared completion, it became apparent that the population chapter could serve as the nucleus of a book that would synthesize the current state of knowledge in climate change, population studies, and related fields for a multidisciplinary audience. O'Neill took the lead in assembling a draft manuscript, MacKellar in revising it, and all three principal authors drafted the concluding Chapter 6. Comments by Professor Thomas Schelling of the University of Maryland and Gordon MacDonald, Director of IIASA, are gratefully acknowledged.

Overall funding for this work was provided by the American Association for the Advancement of Science. MacKellar's involvement was financed in part by the Norwegian Research Council and the Andrew W. Mellon Foundation; O'Neill's work was financed in part by the Wallace Global Fund, the Teresa & H. John Heinz III Charitable Fund of the Heinz Family Foundation, and the William and Flora Hewlett Foundation. Much of O'Neill's work was completed while he was on the staff of the Environmental Defense Fund, New York, NY, USA.

Part I

Chapter 1

Climate Change

The threat of human-induced climate change, popularly known as global warming, presents a difficult challenge to society over the coming decades. The production of so-called "greenhouse gases" (GHGs) as a result of human activity, mainly due to the burning of fossil fuels such as coal, oil, and natural gas, is expected to lead to a generalized warming of the Earth's surface, rising sea levels, and changes in precipitation patterns. The potential impacts of these changes are many and varied – more frequent and intense heat waves, changes in the frequency of droughts and floods, increased coastal flooding, and more damaging storm surges – all with attendant consequences for human health, agriculture, economic activity, biodiversity, and ecosystem functioning.

Because the impacts of climate change are expected to be global and potentially severe, and because energy production from fossil fuels is a fundamental component of the world economy, the stakes in the issue are high. At the same time, a number of aspects of climate change complicate the problem. First, while much is known about the factors governing climate, considerable uncertainty remains in projections of how much climate will change, how severe the impacts will be, and how costly it would be to reduce GHG emissions. Second, because the impacts of today's GHG emissions will be felt for decades into the future, it is not possible to wait and see how severe impacts will turn out to be before taking preventive action. Therefore, if emissions are reduced now, the costs will be borne in the near term while the benefits, which will depend on uncertain projections of future impacts, will be realized largely in the long term. Third, sources of GHG emissions are widely dispersed among nations; no single country could significantly reduce future global climate change just by reducing its own emissions. Even reductions in the more developed nations as a group, which have been responsible for the bulk of historical emissions, would not lead to stabilization of atmospheric concentrations of GHGs. Any solution to the problem must eventually be global.

This chapter briefly outlines the basic science of the greenhouse effect and its role in maintaining the Earth's global climate, the observational evidence for climate change over the past century or more, and the projections of how much and what kind of change may occur over the next 100 years as a result of human activities. It also summarizes the political history of the climate change issue.

1.1   A Climate Primer

The ultimate source of energy driving the Earth's climate is the sun. Sunlight warms the Earth's surface, which responds by re-emitting energy in the form of infrared radiation (heat). If the Earth had no atmosphere, the sun would warm the surface to only about −18 degrees Celsius (°C), a temperature well below the freezing point of water and probably unable to support life. But observations show that the average surface temperature is actually about 15°C. This 33°C difference is due to a natural atmospheric phenomenon known as the greenhouse effect.

1.1.1   The greenhouse effect

A number of gases in the atmosphere are transparent to incoming sunlight but absorb the heat emitted from the Earth's surface and reradiate it in all directions, including back toward the surface. As a result, the surface is warmed more than it would have been in the absence of these gases, which serve a purpose similar to that of the glass walls of a greenhouse.[1]

The greenhouse effect forms the backbone of current understanding of the Earth's climatic history. For example, models of the life cycles of stars indicate that the sun's output has grown by about 30% over the Earth's 4.5-billion-year history. The early Earth therefore received much less solar radiation than it does today. If the atmospheric composition had been the same then as it is now, the planet would have been frozen at the same time that life was beginning to evolve. This apparent impossibility is known as the "faint young sun paradox" (Sagan and Mullen, 1972; Newman and Rood, 1977; Kasting *et al.*, 1988). Currently, the widely accepted solution to this paradox is that the early atmosphere contained much higher concentrations of GHGs than it does today, a proposition for which there is geologic evidence. The additional warming effect these gases provided kept the surface warm enough to allow life to evolve. Since that time, the Earth's broad climatic trends have been regulated in part by processes controlling the atmospheric concentration of GHGs (Walker *et al.*, 1981).

Global warming refers to the enhancement of the natural greenhouse effect due to rising atmospheric concentrations of GHGs resulting from human activities. That the Earth will warm because of the enhanced greenhouse effect is not in dispute.

Debate centers on how much and how fast warming will occur, and how serious the consequences of such changes might be. A small amount of gradual climate change may not be much cause for concern, but if the changes are larger and faster, impacts could be severe. Additionally, small shifts in global mean temperature may prove less important than larger shifts in the frequency and intensity of extreme climate events.

1.1.2 The greenhouse gases

Although GHGs play a critical role in maintaining surface temperature, they make up less than 1% of the atmosphere by volume. The most abundant GHG is water vapor. Although it is responsible for most of the Earth's natural greenhouse effect, its global abundance is not directly influenced by human activity and so it is not considered an anthropogenic (human-generated) GHG. However, its abundance is indirectly affected by anthropogenic emissions in an important way. An increase in the atmospheric concentration of GHGs traps heat near the surface. Part of that heat is expended by evaporating more water, and part warms the surface and lower atmosphere. Assuming relative humidity remains roughly constant, a warmer atmosphere will hold a larger quantity of water, which will be provided by the increased evaporation.[2] Because water vapor is a GHG, this increase in the absolute humidity will further warm the surface and atmosphere, leading to further water evaporation and so on in a positive feedback loop known as the water vapor feedback. The water vapor feedback is a key component of the estimated sensitivity of the climate system to changes in GHG concentrations.

The most important anthropogenic GHGs are carbon dioxide (CO_2), methane (CH_4), nitrous oxide (N_2O), and halocarbons. *Table 1.1* summarizes some key characteristics of these gases.

Carbon Dioxide (CO_2)

Carbon dioxide is, after water vapor, the second most important gas producing the Earth's natural greenhouse effect. Large natural fluxes of CO_2 between the atmosphere and the oceans, land plants, and soils have maintained relatively constant atmospheric concentrations over the past ten thousand years. However, over the past two centuries human activity has released CO_2 at increasing rates. The "pioneer effect" (deforestation in the Northern Hemisphere) was probably the first significant contributor, but with the acceleration of the Industrial Revolution in the 19th century the main source quickly became the burning of fossil fuels such as coal, oil, and natural gas. Over the past several decades, tropical deforestation has also become a significant source, making up about 25% of recent global emissions (Houghton, 1999).

Table 1.1. Summary of main anthropogenic greenhouse gases.

Gas	Concentration (ppb)[a] Preind.	1994	Current growth (%/yr)	Lifetime[b] (years)	Radiative forcing (per molecule, relative to CO_2)	100-year global warming potential (relative to CO_2)
CO_2	~280,000	358,000	0.5	~100 (50–75%)[c] >10^4 (25–50%)	1	1
CH_4	~700	1,720	0.5	12.2	21	21
N_2O	~275	312	0.3	120	206	310
CFC-12[d]	0	0.503	1.4	102	15,600	6,200–7,100

[a] ppb = parts per billion.
[b] Lifetime is defined as the average length of time a present emission will continue to affect atmospheric concentrations.
[c] From O'Neill *et al.* (1997). The atmosphere's response to a CO_2 emission has a distinctly dual nature: at least half the effect of the emission is removed in about 100 years, while the remainder persists for tens of thousands of years or more. The exact fractions and time scales of persistence depend on the assumed future concentration scenario.
[d] CFC-12 is used here as a representative example of the chlorofluorocarbons (CFCs), an important subclass of the halocarbons.
Source: Schimel *et al.* (1996), except as indicated in notes.

Direct measurements of atmospheric CO_2 over the past several decades, as well as measurements obtained from older air trapped in ice sheets, show that CO_2 concentrations have increased about 30% from their preindustrial level (*Figure 1.1*). A number of lines of evidence demonstrate beyond doubt that this rise has been caused by anthropogenic CO_2 emissions, and measurements of the CO_2 content of air bubbles trapped in ice sheets show that the current atmospheric level is higher than at any time in the past 420,000 years (Petit *et al.*, 1999).[3] Once in the atmosphere, CO_2 is not removed by chemical decomposition, but instead is redistributed to other carbon reservoirs. The processes controlling these transfers operate over a wide range of time scales. As a result, one-quarter to one-half of the effect of CO_2 emitted into the atmosphere is removed so slowly that it may be considered essentially permanent, while the remainder is removed in about 100 years (see *Table 1.1*).

Methane (CH_4)

Methane also contributes to the Earth's natural greenhouse effect and is emitted from a range of natural sources, most notably as a product of anaerobic respiration in wetlands. However, current anthropogenic sources are estimated to be about twice as large as natural sources and have led to more than a doubling of preindustrial atmospheric CH_4 concentrations. Methane is released from a wide array of human activities; uncertainty in the magnitude of these fluxes is higher

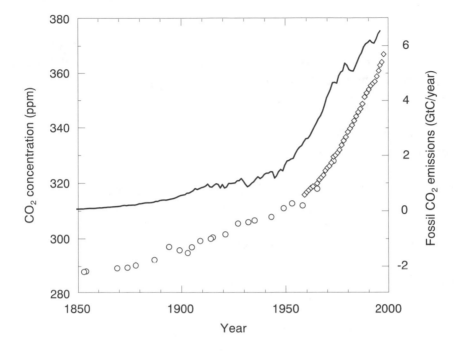

Figure 1.1. Atmospheric CO_2 concentrations (1850–1998) and emissions from fossil fuel use (1850–1996). The continuous curve shows annual CO_2 emissions from fossil fuels in gigatons carbon per year (GtC/year); diamonds are annual means of direct measurements of atmospheric CO_2 in parts per million (ppm); circles are measurements of CO_2 concentrations in air trapped in ice cores. Sources: Keeling, 1994; Neftel *et al.*, 1997; Keeling and Whorf, 1999; Marland *et al.*, 1999.

than uncertainty associated with anthropogenic CO_2 fluxes. Principal sources include livestock such as cattle and sheep (whose digestive systems use fermentation processes that produce CH_4), leakages from natural gas pipelines and coal mines, anaerobic respiration in rice paddies, and biomass burning, with a handful of other sources also contributing (Prather *et al.*, 1995).

Methane has a relatively short atmospheric lifetime of about 12 years; its effect on climate is therefore shorter lived than that of CO_2. It is removed from the atmosphere predominantly by chemical reactions with the hydroxyl radical (OH) and to a lesser extent by soil uptake. The growth rate of CH_4 in the atmosphere has slowed significantly over the past decade (Dlugokencky *et al.*, 1998). While the cause of this decline is uncertain, possibilities include the effect of temperature change (Bekki and Law, 1997), changes in anthropogenic emissions (Law and Nisbet, 1996), and the climatic influence of the 1991 Mt. Pinatubo volcanic eruption.

Nitrous Oxide (N$_2$O)

Nitrous oxide concentrations have increased nearly 15% due to human activity. Sources of N$_2$O have not been well quantified, but the largest natural fluxes to the atmosphere are thought to be from soils (particularly in tropical forests), with a smaller but significant contribution from the oceans. Anthropogenic sources are estimated to be about two-thirds as large as natural sources and are dominated by fluxes from nitrogen-fertilized agricultural fields (Smil, 1999), with additional contributions from a number of other sources including biomass burning, some industrial processes, and cattle and feed lots (Prather *et al.*, 1995). Nitrous oxide is removed from the atmosphere mainly by reactions with sunlight in the stratosphere (upper atmosphere) and has a relatively long lifetime of about 120 years.

Recent attention has focused on the potential for higher emission rates from tropical (relative to temperate) soils following fertilizer use or deposition of NO$_x$ from burning of fossil fuels (Hall and Matson, 1999; Bouwman, 1998). Combined with the anticipation of a further shift in fertilizer and fossil fuel use toward tropical regions, this difference could significantly increase current projections of anthropogenic emissions.

Halocarbons

Halocarbons comprise a number of chlorine-, fluorine-, or bromine-containing gases with generally powerful heat-trapping properties. Halocarbons include chlorofluorocarbons (CFCs) and related compounds such as hydrochlorofluorocarbons (HCFCs), hydrofluorocarbons (HFCs), and perfluorocarbons (PFCs), as well as carbon tetrachloride, sulfur hexafluoride, and methyl chloroform. CFCs have historically been the most important halocarbons in terms of their warming effect. CFCs have no natural sources; they are synthetic compounds used as refrigerants, propellants, blowing agents in the manufacture of foams, and cleaning agents in the production of electronic components. They are essentially inert in the troposphere (lower atmosphere) and therefore have lifetimes of 50 years or more; however, they break down in the stratosphere, where they are the main culprit in stratospheric ozone depletion. Because ozone is itself a GHG, this depletion offsets some of the warming effect of the CFCs, adding a significant uncertainty to the net CFC warming contribution (Schimel *et al.*, 1996). Emissions of CFCs have fallen as the industrialized nations, which are responsible for most of the global total, have complied with the Montreal Protocol on Substances that Deplete the Ozone Layer and its amendments for phasing out production of these chemicals. However, while the HCFCs and HFCs often used as replacements are less efficient ozone depleters, they are still effective GHGs.

Tropospheric Ozone (O_3)

Although about 90% of total ozone is present in the stratosphere, tropospheric ozone (O_3) also has an important effect on climate. This "low-level" ozone is produced through the oxidation of CH_4 as well as through reactions involving a number of precursor gases (e.g., carbon monoxide, nitrogen oxides, and non-methane hydrocarbons), which are themselves produced in part by human activities like biomass burning and fossil fuel combustion. In addition, ozone is transported into the troposphere from the stratosphere. It is destroyed by reaction with ultraviolet light and hydroxyl radicals. Because production and destruction processes vary widely through space and time, ozone concentrations vary with geographical location, altitude, season, and even time of day, making estimation of global trends difficult. Available measurements and modeling studies suggest that tropospheric ozone concentrations in the Northern Hemisphere, where anthropogenic sources are greatest, may have doubled since preindustrial times. Globally averaged, the warming effect of increases in tropospheric ozone levels has probably enhanced the effect due to other GHGs by 10% or more (Stevenson *et al.*, 1998), although effects from ozone have been highly regional.

Radiative Forcing

Because the GHGs have different lifetimes and heat-trapping properties, changes in their abundances affect the Earth's energy balance to different degrees. Any change affecting this balance, whether it is a change in GHG concentrations or a change in the intensity of incoming sunlight, is described as a "radiative forcing." The sixth column of *Table 1.1* shows the relative radiative forcings of the gases on a molecular basis. Molecule for molecule, CH_4, N_2O, and CFCs are much more effective heat-trapping gases than CO_2. The last column gives their 100-year global warming potentials (GWPs), an index that takes into account the different lifetimes of the gases by measuring the cumulative radiative forcing each would contribute over a 100-year period following equal-weight emissions. The values in the table show that, ton for ton, CO_2 emissions will contribute much less to radiative forcing over the next century than will emissions of other GHGs. This comparison should be viewed with some caution since GWPs are subject to a number of uncertainties and are sensitive to, among other things, the choice of time horizon and the future atmospheric concentrations of GHGs (Wuebbles *et al.*, 1995). Nonetheless, they provide an approximate measure of relative radiative forcing effects over a specified time period.[4]

Despite the relatively high GWPs of the other gases, CO_2 is responsible for most of the warming effect resulting from current excess concentrations of all GHGs (see *Table 1.2*). This dominance results from the much greater absolute

Table 1.2. Contribution to radiative forcing due to excess greenhouse gas concentrations.

Gas	1992 radiative forcing[a] (W/m^2)	Percentage of total 1992 greenhouse gas forcing
CO_2	1.56	65
CH_4	0.47	20
N_2O	0.14	6
CFCs and HCFCs	0.15[b]	6
Other	0.03	1

[a] Global average forcing for tropospheric ozone is estimated to be 0.4 watts per square meter (W/m^2), but since its effect is highly regional it is difficult to compare with the well-mixed gases and is not included here.
[b] Includes offsetting negative forcing due to stratospheric ozone depletion.
Source: Schimel *et al.*, 1996.

increase in the atmospheric CO_2 level relative to other gases. For the same reason, its share of responsibility for GHG forcing is expected to grow in the future, although the aggregate warming effect of the other gases is expected to be significant as well.

1.1.3 Other factors influencing climate change

Any climate change induced by the accumulation of anthropogenic GHGs will occur against a background of changes due to a number of other factors. Some of these factors will amplify greenhouse warming; others may offset part of it. Some may, at different times, do both. Each also complicates the task of detecting the anthropogenic component of observed climate change.

Tropospheric Aerosols

Aerosols are small airborne particles that both absorb and reflect radiation and therefore can affect climate directly. They can also affect climate indirectly by altering cloud cover. Aerosols are emitted naturally and by human activity either as particles or as gases that eventually form droplets. Natural sources include soil dust and spray from the oceans; the most important anthropogenic aerosols are sulfates formed from sulfur dioxide gas produced by the combustion of fossil fuels. Incomplete fuel combustion and burning of biomass also contribute significant aerosol fluxes in the form of soot and organic carbon (Jonas *et al.*, 1995).

Different aerosols are thought to have different net effects on climate. Soot from burning of fossil fuels is estimated to exert a positive forcing, while aerosols from burning of biomass and from sulfur emissions are thought to have a cooling effect. Taken together, aerosols are thought to exert a cooling effect on climate and therefore may mask some of the effect of GHGs (Shine and Forster, 1999). Because they generally spend just a few days in the troposphere before returning to the surface in rainfall or through dry deposition, aerosols are not distributed uniformly in the atmosphere but are concentrated near source regions (such as heavily industrialized areas). Their climate effect is therefore regional as well, which makes determining their global contribution to the enhanced greenhouse effect difficult. It is estimated that anthropogenic aerosols may exert a cooling effect that is, on average, about 20% as large as the warming effect of GHGs released through human activity, but in some locations they could offset all of the GHG forcing (Schimel *et al.*, 1996). The uncertainty in this estimate is large, however, and some researchers have argued that because aerosols could play an even larger role, a wider range of future forcing should be considered in climate change scenarios (Hansen *et al.*, 1998a).

Volcanic Activity

Strong volcanic eruptions can inject large amounts of sulfurous gases into the stratosphere, where they are transformed into sulfate aerosols. Because the stratosphere mixes slowly with the troposphere (where aerosols are removed from the atmosphere), stratospheric aerosols disperse globally and persist for several years, reflecting incoming solar radiation and cooling the climate (Minnis *et al.*, 1993). The 1991 eruption of Mount Pinatubo in the Philippines, for example, temporarily reduced global average temperature by several tenths of a degree Celsius. By the end of 1994, however, the aerosols from Pinatubo had been removed from the atmosphere and the dip in temperature had disappeared (see *Figure 1.2* on page 13). Because the effect of volcanos on climate is short lived, future eruptions will not affect long-term climate trends resulting from global warming.

Solar Variations

The sun's output varies slightly over an 11-year cycle associated with sunspots, and presumably over longer time periods as well. Considerable attention has been focused on possible connections between solar cycles and climate, but while intriguing correlations have been found, these efforts have been unsuccessful in proposing a plausible mechanism capable of magnifying the weak magnitude of the solar variations into significant changes in climate (Shine *et al.*, 1995; Kerr, 1996). One possible link is through the influence of changes in solar ultraviolet radiation on stratospheric ozone, which could induce circulation changes in the troposphere large enough to affect surface temperature. However, model estimates of this effect are small (Shindell *et al.*, 1999a). Therefore, although trends in solar output have been an important factor in climate change over geologic history, the sun's influence on climate is expected to be much less important than the impact of rising GHG concentrations in the future.

Natural Variability

Even if climate were not being influenced by any human activities whatsoever, it would still vary from year to year, decade to decade, and even from one century to the next. This natural variability may be caused by external influences such as solar variability or volcanic eruptions, but it may also arise from complicated internal couplings between different parts of the climate system. For example, links between atmospheric and oceanic circulation can produce variations in climate over a time scale of years or decades. A well-known example is the so-called El Niño–Southern Oscillation (ENSO), a periodic change in the temperature of eastern Pacific surface waters that occurs on average every 4.5 years, within a range of 2–10 years, as a result of complex interactions with the atmosphere. El Niño events (the warm phase of ENSO) and their counterparts, La Niña events (the cold phase), are a large source of interannual variability and affect climate around the world, particularly in tropical and subtropical regions (Trenberth, 1997a). The 1997–1998 El Niño was one of the strongest on record, bringing with it a range of regional climate impacts, including severe drought in Guyana, Indonesia, and the Philippines; heavy rains in parts of South America, the west coast of the United States, and Hong Kong; and serious flooding in Kenya.

Although natural variability could amplify or dampen human-driven climate change from year to year or from one decade to the next, it is not likely to reverse the long-term trend toward a warmer climate expected to occur as a result of rising concentrations of GHGs.

1.2   Evidence of Climate Change

The best evidence of climate change over the past century or more – whether natural or human induced – comes from surface temperature records gathered from land- and ship-based weather stations located around the world. These measurements have been compiled into a 130-year record of the globally and annually averaged temperature of the surface of the Earth. Although the trend has not been regionally uniform, *Figure 1.2* shows that average surface temperature has risen by more than half a degree Celsius.[5]

Although climate change comprises changes in a myriad of factors in addition to temperature, such as precipitation, cloud cover, and storm intensity and

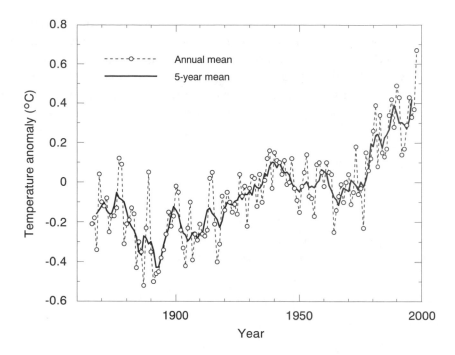

Figure 1.2. Global land–surface air temperature anomaly, 1860–1998, plotted as deviation from 1951–1980 mean. Source: Goddard Institute for Space Studies.

frequency, it is often described only in terms of a change in global average surface temperature. Because it correlates reasonably well with other climate variables, global average temperature is commonly used as a convenient index for climate change more broadly defined. It should be kept in mind, however, that global average measures can be misleading, since it is changes in local or regional conditions that will have direct impact on societies.

Other direct temperature measurements are consistent with the surface temperature record. Upper tropospheric records from balloon ascents show trends similar to those in the surface record since the 1950s. While satellite measurements of tropospheric temperature apparently differ from the surface record, these measurements have been recorded only since 1979 and do not yet provide a long enough record to reliably validate surface temperature trends. Perhaps more important, the two records measure different physical quantities and, when corrected for systematic and transient effects, are not inconsistent (Wentz and Schabel, 1998; Hansen *et al.*, 1995, 1998b; Christy and McNider, 1994). In addition, climate models predict that surface warming should lead to cooling in the stratosphere, and marked stratospheric cooling trends have been observed over the past several decades and are thought to be partly due to surface warming.[6]

A number of lines of indirect evidence for a temperature increase also support the direct observations (Nicholls *et al.*, 1996). Temperature records from un-
derground boreholes, which preserve a record of regional surface temperatures,
are consistent with a temperature rise in many areas. Global sea level has risen
10–25 cm over the past century, and much of this rise is likely due to an increase in
temperature. A short but growing record of precise satellite measurements of sea
level changes promises to reduce uncertainties in the current trend (Nerem, 1999).
Mountain glaciers have retreated, and although trends in sea ice extent and snow
cover are unclear, recent coverage has been below average.

Taken together, this evidence increases confidence in the direct temperature
record. When compared with proxy temperature data for climate over the past
several centuries, the temperature record shows that the 20th century has been the
warmest of any since at least the 15th century and possibly for several thousand
years (Nicholls *et al.*, 1996; Mann *et al.*, 1998, 1999).

Is climate change due to human activity already under way?

The fact that global average temperature has risen over the past century at the
same time that GHG concentrations have increased is not in itself proof that anthropogenic warming is under way. Past climates have displayed variability of
a magnitude similar to historical temperature changes. However, the pattern and
magnitude of the historical temperature rise are consistent with what would be expected to result from historical emissions of GHGs and sulfate aerosols. The inclusion of the cooling effect of aerosols attributable to postwar emissions of sulfur
gases has improved the ability of models to track the historical temperature record
(*Figure 1.3*). Perhaps more important, studies of the detailed spatial and temporal
pattern of the temperature record show that observed geographical, seasonal, and
vertical patterns of temperature change match modeled changes produced by GHG
and aerosol forcing, but not changes that would be expected from volcanic or solar
forcing.

Differentiating anthropogenic temperature change from natural variability
presents a particularly difficult challenge, since human-induced changes may be
expressed through naturally occurring modes of variability (Corti *et al.*, 1999;
Shindell *et al.*, 1999b). However, statistical tests of the significance of recent temperature trends versus background variability indicate that the greenhouse temperature "signal" is beginning to emerge from the "noise" of natural variability (Santer
et al., 1996).

The combined weight of this evidence led the Intergovernmental Panel on Climate Change (IPCC), an international scientific body charged by the United Nations with providing comprehensive surveys of climate change science, to conclude
recently that "the balance of evidence suggests a discernible human influence on

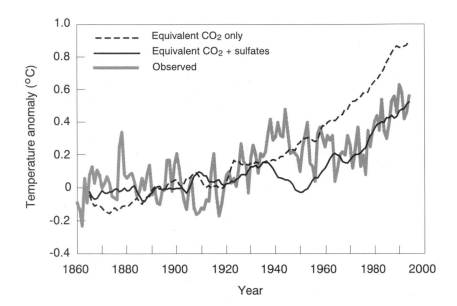

Figure 1.3. Global annual mean warming from 1860–1994 as simulated by a climate model allowing for increases in GHGs only (dashed curve) and GHGs and sulfate aerosols (solid curve), compared with observed temperature change (gray curve). Climate model output is smoothed with 11-year running mean for easier comparison with annual observed data. Temperature values plotted as deviation from 1880–1920 mean. Source: Hadley Centre for Climate Prediction and Research; see Mitchell *et al.*, 1995.

global climate" (Houghton *et al.*, 1996). The question of how much of the observed temperature rise has been due to human influence cannot yet be determined. Over time, detection and attribution of the greenhouse signal are expected to become less ambiguous.

1.3 Contributions to Greenhouse Gas Emissions

Current and historical emissions of GHGs have been dominated by the production of CO_2 from fossil fuel use, primarily in the more developed countries (MDCs). However, when considering other sources of CO_2, as well as emissions of other GHGs, it becomes clear that all regions of the world contribute to emissions of GHGs. *Table 1.3* gives a breakdown of emissions of the three primary GHGs in units of tons of carbon equivalent.[7]

The table shows that, in 1990, 6 gigatons carbon (GtC) equivalent were emitted by MDCs, while less developed countries (LDCs) emitted 4.7 GtC equivalent,

Table 1.3. Anthropogenic greenhouse gas emissions, 1990.

Gas	Total emissions (GtC)[a]		Per capita emissions (tC)[a]		Emissions intensity (tC/10^4 GDP)[a]	
	MDCs	LDCs	MDCs	LDCs	MDCs	LDCs
CO_2	4.59	2.86	3.62	0.71	2.77	8.03
CH_4	0.85	1.15	0.67	0.29	0.51	3.23
N_2O	0.57	0.69	0.45	0.17	0.34	1.94
Total	6.01	4.70	4.75	1.17	3.63	13.20

[a]Emissions in carbon equivalent units calculated using 100-year GWPs (see Note [7]).
Source: Pepper *et al.*, 1992.

making for something on the order of a 55:45 apportionment of current emissions of the three primary gases.[8] Considering CO_2 alone raises the MDC share to over 60%. When measured in terms of contribution to atmospheric concentrations, which integrates the effects of current and past emissions, estimates of the MDC contribution range between 75% (O'Neill, 1996) and 86% (Grübler and Fujii, 1991). Other researchers (Warrick and Rahman, 1992; Grübler and Nakićenović, 1991) have used cumulative historical emissions as a proxy variable for contribution to atmospheric concentrations and found similar results, ranging between 67% and 84%. Part of the variation in results stems from whether emissions from land-use change are considered, since emissions from this source are predominantly from LDCs. In addition, apportioning sinks (such as uptake by forests) as well as emissions can drastically alter results.[9]

The table highlights an additional difference between emission patterns in the two regions: MDCs are high emitters per capita and low emitters per unit gross domestic product (GDP), while the situation is reversed in LDCs. In per capita terms, the average MDC resident emits (so to speak) over four times more GHGs than the average LDC resident. Per capita emissions in MDCs are high because of the high level of economic production per capita. On the other hand, each unit of economic output produced in MDCs contributes to global warming only a little more than one-fourth as much as a unit of economic output produced in LDCs. Emissions per unit GDP are low in MDCs because of a combination of three factors. First, at high levels of economic development, energy intensive industries are less important in overall economic activity than they are in poor countries. Second, those industries that are energy intensive use fuel more efficiently in MDCs than in LDCs. Third, in MDCs there is a preference for cleaner fuels (e.g., nonpolluting natural gas instead of coal). As discussed in Chapter 4, one reason population growth contributes less to rising emissions than one might think is that it is concentrated in the LDCs, where per capita CO_2 emissions are low. On the other hand, LDC economies are expected to grow faster than MDC economies, concentrating growth in the region where CO_2 emissions per unit of GDP are high. Of course, in both cases changes

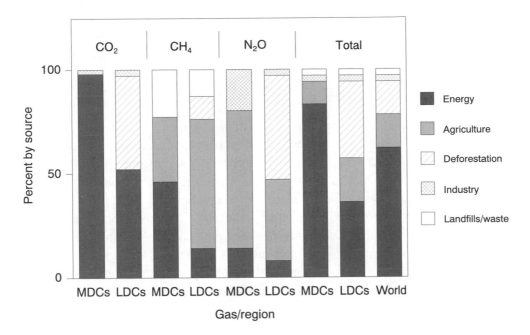

Figure 1.4. MDC and LDC greenhouse gas emissions by source activity, 1990. Totals are in terms of carbon equivalents (see Note [7]). Source: Pepper *et al.*, 1992.

will occur over the coming decades: per capita emissions are likely to grow in LDCs, while emissions per unit GDP are likely to fall as LDC economies undergo structural changes and efficiency improvements.

Data for aggregated regions obscure variation within regions. For example, per capita emissions of CO_2 from energy use are two times higher in North America than in Europe, and two times higher in China and Latin America than in the rest of the developing world. Emissions related to deforestation are not spread evenly throughout the LDCs, but are concentrated heavily in a small number of countries – for example, Brazil. Aggregated data also hide extremes. For example, per capita CO_2 emissions in the United States, one of the better-off and most energy-intensive MDCs, are 200 times higher than in Bangladesh, one of the poorest LDCs. Within countries, there is further heterogeneity between rich and poor and between rural and urban (Murthy *et al.*, 1997).

The sources of GHG emissions also differ between regions. As shown in *Figure 1.4*, the MDCs' contribution to global warming comes mostly from burning fossil fuels. In LDCs, on the other hand, the contribution to emissions splits more evenly into portions due to fossil fuel consumption (36%), agriculture (21%), and deforestation (37%). Worldwide, just over three-fifths (62%) of carbon equivalent emissions are due to energy consumption; changes in land use due to agriculture and deforestation account for another third. Industry (excluding energy consumption) and landfills/waste account for just 6% of total emissions.

1.4   Projecting Future Climate Change

Projecting future climate change begins with projecting GHG emissions, an exercise that must take into account a large number of factors, including population and economic growth, technological change, and energy supply and prices. Because future trends in these factors are highly uncertain, the most common approach is to define a range of scenarios for each. When combined in different ways, these different underlying assumptions produce a range of GHG emission scenarios that are not predictions of the future but instead answer "what if?" questions about possible future emission rates. Models of the GHG cycles are then used to translate the emission projections into a range of future GHG concentrations. In turn, the concentration scenarios are used to drive climate models and produce a range of possible climate effects.

1.4.1   Projections of global mean temperature and sea level rise: The IPCC scenarios

The most prominent projections are based on the IPCC "business as usual" emission scenarios for the 21st century (Leggett *et al.*, 1992), which assume no climate-related policy constraints are placed on GHG emissions. We adopt these scenarios as the basis for our analysis in Chapter 4 because they are the most recent IPCC scenarios available and they are the most widely used basis for analyses of future climate change. The IPCC is currently developing a new set of "no-policy" scenarios, but they were unavailable at the time of this writing. More recent projections by the International Institute for Applied Systems Analysis and the World Energy Council (Nakićenović *et al.*, 1998) provide an updated outlook on energy-related CO_2 emissions. However, as discussed in Chapter 4, we chose the IPCC projections for their more comprehensive treatment of GHGs and more widespread use in climate change studies, which provides a wider context in which to place our results.

Assumptions about underlying factors for the six IPCC scenarios (labeled IS92a–IS92f) are described in *Table 1.4*, and resulting emissions of CO_2 and CH_4 are shown in *Figure 1.5*.

The two extreme scenarios are IS92c and IS92e. In scenario IS92c, slow economic growth is coupled with a low population projection, a pairing that is probably unlikely given that – as discussed in Chapter 2 – more rapid socioeconomic development is one (although not the only) element associated with more rapid fertility

Table 1.4. Summary of assumptions in the IPCC IS92 emission scenarios.

Scenario	Population	Economic growth (%/yr)	Energy supplies
IS92a,b	World Bank 1991: 11.3 billion by 2100	1990–2025: 2.9 1990–2100: 2.3	12,000 EJ conventional oil; 13,000 EJ natural gas; solar costs fall to US$0.075/kWh; 191 EJ of biofuels available at US$70/barrel
IS92c	UN medium-low case: 6.4 billion by 2100	1990–2025: 2.0 1990–2100: 1.2	8,000 EJ conventional oil; 7,300 EJ natural gas; nuclear costs decline by 0.4%/yr
IS92d	UN medium-low case: 6.4 billion by 2100	1990–2025: 2.7 1990–2100: 2.0	Oil and gas same as IS92c; solar costs fall to US$0.065/kWh; 272 EJ of biofuels available at US$50/barrel
IS92e	World Bank 1991: 11.3 billion by 2100	1990–2025: 3.5 1990–2100: 3.0	18,400 EJ conventional oil; gas same as IS92a,b; phaseout of nuclear by 2075
IS92f	UN medium-high case: 17.6 billion by 2100	1990–2025: 2.9 1990–2100: 2.3	Oil and gas same as IS92e; solar costs fall to US$0.083/kWh; nuclear costs increase to US$0.09/kWh

Abbreviations: EJ = exajoule; kWh = kilowatt hours.
Source: Leggett *et al.*, 1992.

decline. The scenario also assumes an expansion of nuclear and renewable energy sources. Resulting CO_2 and CH_4 emissions peak in the early to mid-21st century at levels about 20% higher than current levels and decline thereafter. In contrast, scenario IS92e assumes rapid economic development, a medium population scenario, and continued reliance on fossil fuels. Emissions of CO_2 increase by a factor of five over the next century, while CH_4 emissions double. Uncertainty in unconstrained future emissions is large.

The IS92 scenarios also project a distinct shift in the source of emissions. For example, under IS92a, CO_2 emissions from LDCs equal those from MDCs by 2025–2035. Accounting for emissions of CH_4 and N_2O as well, the crossover date is 2015. By 2100, emissions of the three gases from countries currently defined as LDCs are twice those of the MDCs.

The IPCC used the IS92 scenarios for CO_2 and CH_4, combined with corresponding ranges of emissions for other GHGs and sulfate aerosols, to investigate future climate change. Using GHG cycle models and simple climate models, they project increases in global average temperature by the year 2100 of 1–3.5°C, with a central estimate derived from scenario IS92a of about 2°C.[10] Roughly half this

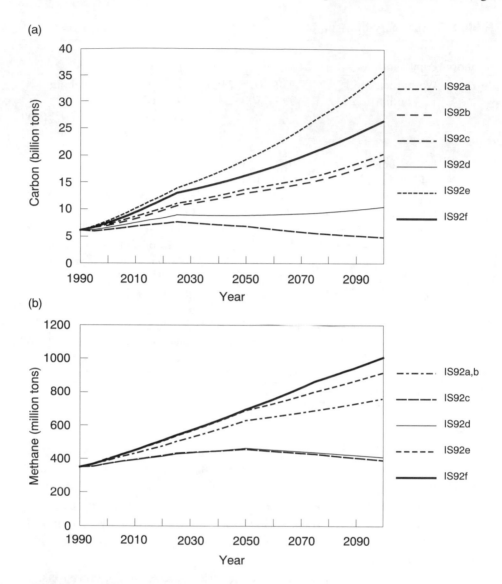

Figure 1.5. Projections of future anthropogenic (a) CO_2 and (b) CH_4 emissions according to the IPCC IS92 emission scenarios, 1990–2100.

range of uncertainty is attributable to the uncertainty in future emissions and the other half to uncertainty in the sensitivity of the temperature response to increases in GHG concentrations. Because of inertia in the climate system, this warming represents only 50–90% of the total warming embodied in the projected emissions in these scenarios. To get an idea of the degree of inertia, an instantaneous increase in atmospheric GHG concentrations would not be fully reflected in the climate for several decades, due mainly to storage of heat in the oceans. Thus, even if emissions were to decline after 2100, global mean temperature would continue to rise.

It might be thought that a world in which the average temperature is a few degrees warmer would not differ greatly from the world in which we now live. However, average surface temperature is an index of the overall state of the Earth's climate, and a change of a few degrees is a major global event. For example, during the last ice age, global average temperature was probably just 5°C lower than it is today, but this difference was enough to cause ice sheets to cover vast areas of Europe and North America that are currently ice free. The high end of the range of IPCC projections, especially beyond 2100, approaches a warming of a magnitude similar to that of the glacial cooling.

In addition, all scenarios produce a global warming that is probably more rapid than at any time in the past ten thousand years (Nicholls *et al.*, 1996). The climate changes actually experienced in particular regions over particular decades are expected to vary considerably. Natural variability is likely to affect the rate of warming in unpredictable ways, and modelers are more confident of trends at the scale of continents or hemispheres than they are of regional or local effects. Nonetheless, some broad characteristics consistently emerge from most modeling studies. For example, warming tends to be greater over land than over the ocean, and greater toward the poles than near the equator. In addition, most of the warming in high northern latitudes occurs in winter, and differences between daytime and nighttime temperatures are reduced in most regions in all seasons. Future sulfur emissions will also influence regional climate change due to the cooling effect of sulfate aerosols; in some regions, such as parts of Asia, aerosol cooling effects could even outweigh greenhouse warming.

Global mean sea level is expected to rise as the climate warms, mainly due to the thermal expansion of seawater, with an additional contribution from the melting of glaciers and ice caps. Taking into account the projected range of GHG emissions in the IS92 scenarios, as well as model uncertainties, the range of projected global mean sea level rise by 2100 is 15–95 cm (Warrick *et al.*, 1996), with a central estimate of 50 cm corresponding to the mid-range IS92a scenario. Contributions from the Greenland and Antarctic ice sheets are currently expected to be minor, but there is considerable uncertainty in this regard and the possibility of substantial sea level rise in the long term from melting of the West Antarctic Ice Sheet cannot be ruled out (Oppenheimer, 1998). Sea level rise at particular locations could vary widely from the global mean as a result of regional differences in both oceanic heating and circulation, and local land movements, which can be due to long-term geophysical processes or to human influences such as draining of wetlands, groundwater pumping, or sedimentation. Even if GHG concentrations are stabilized by 2100, sea level will continue to rise for several centuries due to lags in the response of oceans and ice sheets (Wigley, 1995).

1.4.2   Projected changes in other climate-related factors

Projections of global temperature trends are considered to be more reliable than projections of any other climatological variable. However, there are a number of general characteristics of changes in other variables on which there is basic agreement (Kattenberg *et al.*, 1996).

For example, all general circulation models, or GCMs (see *Box 1.1*), predict an increase in global average precipitation. In general, higher temperatures lead to more evapotranspiration balanced on average by more precipitation, and there are signs that the global hydrological cycle is already intensifying (Karl and Häberli, 1998).[11] But changes are not expected to be regionally or seasonally uniform. Some areas may experience decreases in precipitation, and the seasonal timing of precipitation may change. As a result, in some regions droughts and/or floods may be more severe while in others they may be less severe.

Soil moisture projections are probably more useful than those for precipitation, since they reflect changes in precipitation, evapotranspiration, and runoff throughout the year. Most models project wetter soil conditions during winter months in high-latitude regions and drier conditions during summer months in mid-latitude regions (in southern Europe and North America, for example). However, the results are highly uncertain and are sensitive to whether sulfate aerosols (which have regionally specific climate effects) are included in the projections.

1.4.3   Extreme weather events

Some kinds of extreme weather are likely to be more frequent. For example, most models show an increased number of extremely hot days (and fewer extremely cold days). Many project more frequent heavy rainfalls (e.g., Hennessy *et al.*, 1997) and at the same time more frequent and severe droughts. Changes in historical trends in precipitation extremes have been identified in some recent studies. For example, Dai *et al.* (1998) present evidence of a greater percentage of global land area experiencing either severe drought or moisture surplus, although the current fraction is not outside historical experience. In addition, the proportion of total precipitation falling in extreme events has increased significantly in the United States, where an extensive, high-quality dataset is available (Karl and Knight, 1998). Smaller increases have been documented in many other regions of the world (Karl and Häberli, 1998). However, some areas show no trend, so that while results are suggestive, no strong conclusions at the global level can yet be drawn (Nicholls *et al.*, 1996).

Box 1.1. General circulation models.

General circulation models (GCMs) are computer models based on physical laws and empirical relationships that are used to simulate and project climate. Some models describe only the atmosphere or the oceans; other "coupled" models encompass both. Land surface processes are also included in some GCMs. GCMs represent the climate system as a three-dimensional grid with a typical resolution of about 250 km horizontally and 1 km vertically. Equations of motion, heat, and radiative transfer are solved at successive intervals of time to track variables such as temperature, pressure, and humidity at each grid point. Smaller-scale processes – related, for example, to clouds or small-scale mixing in the oceans and atmosphere – are represented in a simplified manner.

Typical model experiments begin by simulating current climate as a control. Current coupled models can successfully reproduce the large-scale features of the climate, including seasonal surface temperatures and precipitation rates (Gates *et al.*, 1996). They are less effective when reproducing regional features, and differences between models become greater for smaller-scale features. "Equilibrium" experiments, often using models that employ a simplified representation of the oceans, are performed by setting GHG concentrations in the model at a higher level (double the preindustrial concentration of CO_2 is a standard benchmark) and allowing the model to fully adjust to a new equilibrium. The new equilibrium temperature, precipitation, and other climatic variables are taken as a projection of conditions under higher GHG concentrations. "Transient" (as opposed to equilibrium) experiments are performed by adding GHGs to a coupled model in a realistic time pattern instead of instantaneously increasing them. Although computationally more demanding, this kind of model experiment provides insights into the kinds and rates of changes that might occur as GHG levels increase.

Recent advances have addressed the tendency of coupled atmosphere–ocean models to drift away from the control climate even in the absence of external forcing. This drift is often counteracted by adjusting exchanges of heat and fresh water across the ocean–atmosphere boundary to maintain equilibrium, a technique known as "flux adjustment" that has long been viewed as indicative of a shortcoming of climate models (Kerr, 1997; Trenberth, 1997b). Improvements in the representation of sub-grid-scale processes have facilitated the development of models that do not require flux adjustment (Gregory and Mitchell, 1997; Bryan, 1998; Carson, 1999), increasing confidence in results.

Projections cannot be "tested" against observations. As an alternative, GCM simulations of current climate are sometimes compared with observations that have been processed by models used to make short-term weather predictions. GCMs have also been used to reproduce climates that existed at different times in the Earth's history under different conditions as a means of increasing confidence in model results.

El Niño-Southern Oscillation (ENSO) events (see Section 1.1.3) are a major source of variability in precipitation and temperature in many parts of the world. Whether climate change will lead to increased frequency, duration, or intensity of El Niño or La Niña events is a central research question that has received increasing attention in part because of the persistent El Niño conditions of 1990–1995 and the severe El Niño event of 1997–1998, which was one of the strongest on record (McPhaden, 1999). It is being addressed through both statistical analyses of the historical record and climate model simulations of future conditions. While results are suggestive of a link between climate change and more frequent or persistent events, neither strategy has yet produced definitive answers.

Trenberth and Hoar (1996) have argued that a trend toward more frequent El Niño events since the mid-1970s and the magnitude and persistence of El Niño conditions during the first half of the 1990s were highly unusual relative to the 120-year historical record. However, their work has been disputed (and, in response, defended) on statistical grounds (Rajagopalan *et al.*, 1997; Trenberth and Hoar, 1997), and separate analyses have found that persistent El Niño conditions are not unusual, especially when paleoclimate data extending beyond the historical record are added to the analysis (Allan and D'Arrigo, 1999). Nor have GCM simulations produced consistent results. A number of studies have found little evidence of significant changes in El Niño events under simulated global warming scenarios (Tett, 1995; Knutson *et al.*, 1997; Smith *et al.*, 1997), but a study using a model with better resolution of tropical processes important to simulating El Niño events found an increase in El Niño–like conditions as a result of increased greenhouse gas forcing of climate (Timmerman *et al.*, 1999).

Modeling studies are inconclusive on whether severe storms such as hurricanes or monsoons might change in frequency, intensity, or time or area of occurrence (Kattenberg *et al.*, 1996). Tropical cyclones are not adequately simulated in GCMs, and since the natural variability of such storms is large, small trends are not likely to be detectable. Regional simulations lend some support to theories that cyclone intensities could increase in a warmer world (Knutson *et al.*, 1998). On the other hand, the limited available evidence suggests that climate change will have little or no effect on global frequency of tropical cyclones, although local and regional frequencies could change significantly if other factors that affect the formation of cyclones and the paths they follow, such as El Niño events, are themselves affected by climate change (Henderson-Sellers *et al.*, 1998).

1.5   Impacts on Society and Ecosystems

Climate change is expected to have a wide range of direct and indirect impacts on human health, ecological systems, and socioeconomic sectors. Both the magnitude and rate of change, as well as possible changes in climate variability, will be important in determining impacts. Many impacts will be adverse, and some, like species loss or ecosystem changes, may be irreversible. Others may be beneficial. The ultimate effects of climate change on socioeconomic or ecological systems will depend on three broad factors: the characteristics of the change in climate, the sensitivity of the system to a given change, and the capacity of the system to adapt to climate change. The vulnerability of a particular society is generally defined by the combination of sensitivity and adaptability. Those societies or sectors that are especially sensitive or are least able to adapt are the most vulnerable.

Projecting impacts, especially at the regional or local level, involves a high degree of uncertainty, and quantification is difficult. In addition, studies to date have had a limited scope: most have been carried out assuming a background climate change corresponding to an increase in GHG concentrations equivalent to a doubling of CO_2 concentrations (projected to occur in the early to mid-21st century). Much less is known about transient impacts while climate is changing, or about impacts for more extensive climate changes. Moreover, human capacity to adapt makes it difficult to project the ultimate effect of impacts on human societies. Successful adaptation will depend on, *inter alia*, access to knowledge and technology, institutional arrangements, and availability of financing. Because these factors are likely to differ significantly across populations, regional effects are highly uncertain.

In Chapter 5, we discuss the expected impacts of climate change on agriculture, human health, and environmental security, focusing in particular on vulnerable populations and on the question of whether rapidly growing populations are more vulnerable to the impacts of climate change than are more slowly growing ones. We therefore reserve discussion of impacts in these important areas for later treatment. Here, we briefly outline a number of other impacts that highlight the sensitivity of some systems to the rate and magnitude of climate change, and the range of response options available to societies. We also discuss potential differences in impacts at the regional level.

1.5.1 Ecosystems

The projections of climate change discussed in the previous section involve rates of change in excess of the rate at which many ecosystems can adapt. Combined with the influence of higher CO_2 levels and changes in soil characteristics, fires, pests, and diseases, all of which are likely to selectively favor particular species, this suggests that climate change has the potential to change the location, functioning, and species makeup of many ecological systems (Root and Schneider, 1995). Some adaptation options are available, such as establishment of migration corridors, restoration efforts, and changes in land-use practices. However, the existence of additional stresses from habitat fragmentation and pollution, combined with the expected pace of climate change, make ecosystems vulnerable. For example, doubled equivalent CO_2 scenarios suggest that a global average of one-third of forested areas would experience changes in broad vegetation type (Kirschbaum and Fischlin, 1996). Changes would vary by region, with the greatest impacts being felt at higher latitudes, where as much as two-thirds of forested areas could change type. Aquatic species would also be affected. Changes in water temperature, as well as in flow variability caused by changes in precipitation patterns, would lead to extinction of some species and shifts in the ranges of others. Some coastal ecosystems such as saltwater marshes, coastal wetlands, and coral reefs are especially vulnerable to the combined effects of sea level rise, changes in climate, and possible changes in the erosion of shores.

1.5.2   Water

Water resources in many regions of the world are likely to be under increasing stress over the next century, particularly in developing countries with growing populations and economies. Climate change may have adverse or beneficial impacts on water systems. Adverse impacts could worsen flooding in temperate regions and increase scarcity in many other regions. Impacts are expected to be greatest in those areas that are already under stress, particularly arid or semiarid regions in LDCs, and the most vulnerable populations are those in dry regions that rely on single sources such as wells or isolated reservoirs (Kaczmarek, 1996). In one global study (Strzepek *et al.*, 1995) based on a GCM simulation of climate change, water scarcity in 2025 was found to be eased in some areas but exacerbated in others. Results must be interpreted with caution, however, because outcomes for particular countries and regions depend heavily on the particular model used to generate the climate change scenario (Kaczmarek *et al.*, 1995).

Climate change is likely to affect water demand as well as supply. For example, a drier climate may induce an expansion of irrigated agriculture, leading to demand increases above even the rising needs of growing populations and economies. The combination of increasing demand, sensitivity to fluctuations in precipitation, and limited resources available for adaptation makes many regions vulnerable to impacts on water systems. A range of adaptive avenues is available, including increasing efficiency of water systems (particularly with regard to agricultural water use), expanding system capacity, changing pricing practices, developing institutions for managing water, and improving urban planning.

1.5.3   Sea level rise and coastal systems

Sea level rise is expected to threaten coastal systems at the same time that population in these areas is growing faster than average national rates. The most dramatic effect of sea level rise is inundation, that is, the loss of land area to the sea. The populations most at risk are those living in small island-states (Pernetta, 1992; Bijlsma, 1996), some of which, like the Marshall Islands, may almost completely disappear as a result of sea level rise. Many low-lying continental coastal areas are also susceptible. If countermeasures are not taken, these areas will gradually be lost to human habitation. If countermeasures are taken, the effects of sea level rise can be mitigated to some degree; the price will be forgone investment in other areas. Societies can respond to sea level rise through retreat, accommodation, or protection, or, more likely, some combination of the three.

Concern has been focused on the inundation of densely populated low-lying areas such as the Nile Delta and coastal cities. Because the value of land is so high in such regions, these areas are most likely to pursue protection strategies, although required responses may be daunting. For example, preserving the Nile Delta region would require coastal protection structures on the scale of the Netherlands' Great Delta Works, along with comprehensive management of the water supply through wastewater recycling systems, increased groundwater exploitation, and regulation of water drawn from the Nile (Stanley and Warne, 1993). The response to sea level rise in areas of low population density, in contrast, will likely be retreat or accommodation.

Sea level rise has a range of impacts less profound than inundation, such as intrusion of seawater into aquifers and increases in the frequency of flooding of low-lying coastal areas. The severity of damage will depend on local conditions in addition to the extent of sea level rise (Smil, 1990), but some general conclusions have been drawn regarding global risk. According to current estimates, in an average year about 46 million people worldwide experience flooding due to storm surges. If the sea level rises 50 cm – the best guess for the year 2100 under the IPCC's central IS92a scenario – that number will likely double (Bijlsma, 1996). Local conditions will make some regions, including the Indian Ocean coast, the southern Mediterranean coast, the African Atlantic coast, and Caribbean coasts, more susceptible than others. These results do not account for population growth, the current trend toward disproportionate growth of coastal populations, possible changes in storm frequency, or potential adaptation measures. Clearly, coastal regions will be affected by a confluence of climatic, demographic, and socioeconomic forces.

1.5.4 Natural catastrophes

The 1990s set a record for losses from natural disasters caused by extreme weather events. Floods in China and Central Europe, Hurricane Andrew in Florida, and Hurricane Mitch in Central America are the most recent highlights of a general upward trend in disaster losses since at least the 1960s. Over this period, economic damages from all natural catastrophes (including events unrelated to weather, such as earthquakes and volcanoes) increased by a factor of 9, and the cost to the insurance industry increased by a factor of 15 (Munich Re, 1997). Most recently, the 1997–1998 El Niño event caused an estimated US$33 billion in damage and cost 23,000 lives worldwide (Kerr, 1999).

The principal cause of this long-term trend toward increased losses has been the growth of population, capital, and infrastructure in catastrophe-prone areas – flood basins of major river systems, low-lying coastal areas, earthquake fault zones, and areas exposed to hurricane risk. Many in the insurance industry have adopted the view that a trend toward increased severity and frequency of extreme events has played a role as well, but no consensus exists on this matter (Dlugolecki, 1996). The role of changes in the frequency of extreme weather events (see Section 1.4.3) is difficult to determine; data showing an increase in the frequency of such events in some regions are suggestive of a general trend, but no definitive conclusions can yet be drawn at the global scale.

If extreme events become more frequent and intense as climate changes in the future, damages may escalate as a result of synergistic effects between extreme weather and concentrations of population and economic value in vulnerable areas. This potential was clearly illustrated by the devastating consequences of the severe weather associated with the 1997–1998 El Niño event, although whether such events will become more common as a result of climate change remains an open question (see Section 1.4.3). At the same time, adaptive measures will be possible. For example, as consumers are forced to shoulder more of the burden of catastrophic damages, new forms of insurance and other risk-transfer mechanisms may come into play (Stripple, 1998).

1.5.5   Surprise events

Most modeling studies of future climate can only produce relatively smooth, gradual changes because of the nature of the models employed. However, studies of past climates show that the Earth's climate system is capable of undergoing large, sudden shifts. During the most recent ice age (beginning about one hundred thousand years ago and ending about ten thousand years ago), climate underwent a number of repeated fluctuations; temperatures in the Northern Hemisphere sometimes increased by over 5°C in as little as a few years to a few decades (Alley *et al.*, 1993). Equally large and rapid changes took place as the planet began warming to its present climate at the end of the last ice age just over ten thousand years ago (Dansgaard *et al.*, 1989a). Recent evidence from ice cores shows that temperatures in Greenland warmed by 5–10°C in a few decades or less (Severinghaus *et al.*,

1998). Data from Antarctic ice cores indicate associated changes extended to the Southern Hemisphere as well (Blunier *et al.*, 1998; Steig *et al.*, 1998).

The possibility of triggering large, sudden events is an important consideration in weighing the potential consequences of anthropogenic climate change (Broecker, 1987). The abrupt changes during the ice ages were probably driven by sudden re-organizations of oceanic circulation, particularly the shutdown and restarting of the large-scale overturning of ocean waters known as thermohaline circulation, or the oceanic "conveyor belt." This circulation pattern is driven by the sinking of cold, salty (and therefore dense) water in the North Atlantic and delivers a large amount of heat from the Tropics to the North Atlantic region, maintaining winter tempera-tures in northern Europe at levels considerably higher than they would be otherwise. It likely has multiple equilibrium states, however, and its apparent ability to switch rapidly from one to another has led it to be called the Achilles' heel of the climate system (Broecker, 1997).

Studies of future climate change using coupled ocean–atmosphere models show that changes in temperature and precipitation attendant upon global warming could lead to a slowdown or complete shutdown of the conveyor belt. One study based on an improved model not requiring artificial "flux adjustment" (see *Box 1.1*) projects a 25% decrease in ocean circulation that could begin within the next several decades (Wood *et al.*, 1999). Other studies suggest that beyond particular threshold warm-ing levels, circulation may collapse and not recover (Manabe and Stouffer, 1994) and that the threshold levels and potential for recovery may depend on the rate of warming (Stocker and Schmittner, 1997).

Projecting changes in the strength of the thermohaline circulation is compli-cated by its sensitivity to changes in regional temperature and precipitation, which affect the density of North Atlantic surface waters and therefore the rate of deep wa-ter formation. Model projections of regional precipitation changes are not in good agreement (Rahmstorf, 1999) and are made even more difficult by their dependence on the North Atlantic Oscillation (NAO), a periodic change in the atmospheric pres-sure difference along a North–South transect of the Atlantic Ocean that accounts for a significant fraction of interannual climate variability in the region (Hurrell, 1995). The NAO has been in an unusually high phase since the mid-1970s, and recent observations showing a temporary but pronounced slowing of one compo-nent of deep water formation have been tentatively linked to the extreme state of the NAO (Dickson *et al.*, 1999).

The consequences of significantly altering the thermohaline circulation could be serious, at least in regions surrounding the North Atlantic where, counterintu-itively, climate could cool dramatically. Impacts on other regions are difficult to predict; past reorganizations of ocean circulation produced global impacts, but re-liable estimates of the combined effects of global warming and circulation changes are beyond current modeling capabilities (Rahmstorf, 1997). Identifying thresholds beyond which additional climate change will greatly increase the risk of potentially catastrophic changes remains an important but elusive goal.

The possibility of unstable feedback involving climate change has also been raised. For example, large quantities of CH_4 are trapped in ice crystals in ocean sediments and in tundra soils – in general known as methane clathrates. Although at present the likelihood of destabilizing these clathrates on a large scale is considered to be low, it is possible that rising temperatures could begin to melt them. Methane would then be released to the atmosphere, further raising temperatures, which would melt more clathrates and release more CH_4, and so on in a positive feedback loop (Lashof, 1989).

Other potential feedbacks have been identified; however, because the climate system is complicated, it may not be possible to anticipate all of them. The history of the ozone issue provides a particularly relevant example. Until the mid-1980s, most projections of ozone depletion due to the release of CFCs indicated little potential for serious depletion (US National Research Council, 1984). The discovery in 1985 of the "ozone hole" (Farman *et al.*, 1985), an area of near total stratospheric ozone loss over Antarctica that occurs in the Southern Hemisphere spring, came as a complete surprise. The models did not account for the kinds of chemical reactions that led to the formation of the hole and so could not possibly have predicted its occurrence. It is conceivable that the 21st century could produce analogous events related to global warming.

1.6   Global Warming: A Historical Sketch

Although global warming is relatively new as an issue of major international concern, it has received the attention of scientists for over a century.[12] The greenhouse effect was given its name in 1827 by the French mathematician Jean-Baptiste Fourier, who first compared the atmosphere to the glass walls of a greenhouse. A few decades later, British physicist John Tyndall identified the main GHGs. He also suggested that changes in GHG concentrations might help explain the swings in climate known to have taken place during the ice ages. But it was not until 1896 that Swedish chemist Svante Arrhenius first connected rising combustion of coal and its associated release of CO_2 with the potential to alter the current climate (Arrhenius, 1896).

During the first half of the 20th century little attention was paid to the potential for global warming, mainly because few scientists thought CO_2 was accumulating in the atmosphere. Although George Callendar compiled existing measurements in the 1930s and concluded that some significant buildup had occurred (Callendar, 1938), most thought the measurements too imprecise to be given much weight and assumed instead that the oceans, with their vast capacity to store carbon, would absorb any emissions from human activity. But in 1957, Roger Revelle and Hans Suess, in a landmark paper (Revelle and Suess, 1957), showed that although the oceans can store enormous amounts of carbon, the chemistry of seawater creates a bottleneck to absorption of CO_2 from the atmosphere. This property, which has come to be known as the Revelle effect, meant that continued emissions of CO_2 could cause significant accumulation in the atmosphere and potentially alter climate, leading the authors to the often-quoted observation that "human beings are now carrying out a large scale geophysical experiment that could not have happened in the past nor be reproduced in the future."

Over the next two decades, precise measurements of atmospheric CO_2 made by Charles Keeling showed conclusively that CO_2 was indeed accumulating (see *Figure 1.1*). Although the period also saw the development of the first generation of climate models and GCMs for projecting climate change (Manabe and Wetherald, 1967, 1975), widespread concern was restrained by the lack of observed warming. In the early 1970s, for example, a string of colder-than-normal years focused some attention on the possibility of global cooling due to natural ice age cycles.

However in the late 1970s global temperatures resumed their rise. Climate models became more sophisticated, increasing confidence in their projections, which were shown to be broadly consistent with the 20th century's generally rising temperature trend. In addition, in the mid-1980s the contribution of GHGs other than CO_2 was fully appreciated (Ramanathan *et al.*, 1985), leading to a rough doubling of the perceived magnitude of the problem. Meanwhile, the decade of the 1980s was on its way to becoming the warmest on record.

It was during this decade that global climate change gained momentum as a political issue (Bodansky, 1993). At the First World Climate Conference in 1979 there was discussion of the problem but no call for action to curb emissions of GHGs. However, in 1985 at a conference in Villach, Austria, scientists concluded that significant climate change was likely and issued the first call to scientists and governments to begin exploring alternative policies. That same year, the public profile of the issue was raised by the discovery of the ozone hole and subsequent work linking it to CFCs, which highlighted the threat of dangerous anthropogenic impacts on the atmosphere. The 1988 heat wave and drought in the United States brought the climate issue to public attention even more forcefully. At an influential conference in Toronto that year, the participants (largely environmentalists) called for 20% cuts in global CO_2 emissions by 2005, to be funded mainly by the industrialized countries.

Also in 1988, at the request of several governments (primarily the USA), the Intergovernmental Panel on Climate Change (IPCC) was formed by the World Meteorological Organization and the United Nations Environment Programme. Its mandate was to assess the scientific basis for climate change and its potential impacts, and to evaluate potential response strategies. It issued its first comprehensive report in 1990 (Houghton *et al.*, 1990), projecting significant climate change under "business as usual" conditions. The report quickly became accepted as an authoritative statement on the issue. In the same year, the Second World Climate Conference was held in Geneva, Switzerland. It was attended by both scientists and government representatives, and for the first time included significant participation from developing countries. The conference declaration applauded the goals of some countries to stabilize GHG emissions at 1990 levels by the year 2000, but did not go so far as to recommend adoption of specific targets by developed countries in general.

By this time there was a clear demand for an international agreement on global warming, and a negotiation process was begun through the UN to produce a convention on climate change. Remarkably, the Framework Convention on Climate Change (FCCC) was produced in less than 18 months, despite the remaining scientific uncertainties, the world's heavy dependence on fossil fuels, and the divergent interests of the nations involved (Bodansky, 1993). The FCCC calls for the "stabilization of GHG concentrations in the atmosphere at a level that would prevent dangerous anthropogenic interference with the climate system." It stipulates that parties should be guided by a number of principles in achieving this objective, including equity (specifically, that developed countries should "take the lead" in climate change mitigation), the precautionary principle (when in doubt, to err on the side of caution), cost-effectiveness, and sustainable development. The FCCC was signed by representatives of 165 governments at the United Nations Conference on Environment and Development (the "Earth Summit") in Rio de Janeiro in 1992. It has now been signed and ratified by 176 countries. It legally came into force in March 1994.

The FCCC suggests that industrialized countries aim to stabilize their emissions at 1990 levels by the year 2000. This goal is likely to be met in the aggregate (United Nations, 1998b), mainly because emissions from Russia and East European countries fell sharply in the early 1990s due to economic crises and are only beginning to recover. With a few exceptions (notably France, Germany, and the United Kingdom), other countries do not anticipate achieving the FCCC goal. Furthermore, aggregate emissions are expected to grow again beyond 2000.

As discussed below, stabilizing emissions of GHGs will not lead to a stabilization of atmospheric concentrations, the goal of the convention; deeper cuts in emissions would be required to meet this goal. In recognition of these shortcomings, in 1995 parties to the FCCC agreed that further efforts would be needed to achieve the convention's objective and signed the "Berlin Mandate" to negotiate binding commitments toward stabilizing GHG concentrations. Negotiations were undertaken, leading to the signing of the Kyoto Protocol in 1997 by representatives of over 150 governments. If the protocol is ratified, it will require industrialized countries to reduce aggregate emissions of six GHGs – CO_2, CH_4, N_2O, HFCs, PFCs, and SF_6 (sulfur hexafluoride) – by about 5% relative to 1990 emission levels. This goal is to be achieved over the compliance period 2008–2012. The protocol envisions a system of emissions trading in which countries can make deeper reductions than required and sell excess allowances or emit at rates above their cap and buy allowances for the difference. Individual countries accepted different caps; Australia, for example, is allowed to increase emissions 8% above 1990 levels, while the United States and the European Union must decrease emissions by 7% and 8%, respectively. The aim of such a system is to reduce the overall costs of emission reductions (Skea, 1999). In principle, a country in which domestic reductions would be expensive could reduce costs by "buying" (in other words, financing the achievement of) cheaper reductions elsewhere. In practice, there are a host of issues to be addressed before such a system could be made operational, including questions of monitoring and verification, scientific uncertainties in accounting for sources and sinks related to land-use change (Jonas *et al.*, 1999), and concerns about whether cost reductions will actually be realized even with an international trading system in place (Hahn and Stavins, 1999).

Developing countries have no new commitments under the protocol, but can volunteer to accept binding emission caps. They can also participate in emission reductions through the Clean Development Mechanism (CDM), which would allow industrialized countries to fund, and receive credit for, emission reduction projects in developing countries. A number of stumbling blocks remain in making the CDM operational such that projects could contribute significantly to achieving emission reduction goals, including how to define baseline emissions that would have occurred in the absence of the project, how to distribute benefits between donor and host countries, and how to report and verify results (Trexler and Kosloff, 1998).

International negotiations are continuing with the aim of fleshing out the details of the Kyoto agreements. A conference in Buenos Aires in 1998 concluded with an agreement to produce more specific guidelines in 2000. In the longer term, the protocol calls for periodic assessments of the climate issue and negotiation of emission limits for a series of compliance periods stretching beyond 2012.

1.6.1 The road ahead: Stabilizing GHG concentrations?

Achieving the goal of stabilizing GHG concentrations at a level that would prevent dangerous interference with the climate system will first require determining what amount of climate change would constitute dangerous interference, a task complicated by uncertainty in both the climate response to GHG forcing and the capacity of social, economic, and ecologic systems to adapt. It will also require a subjective evaluation of the kinds of changes that may or may not be tolerable.

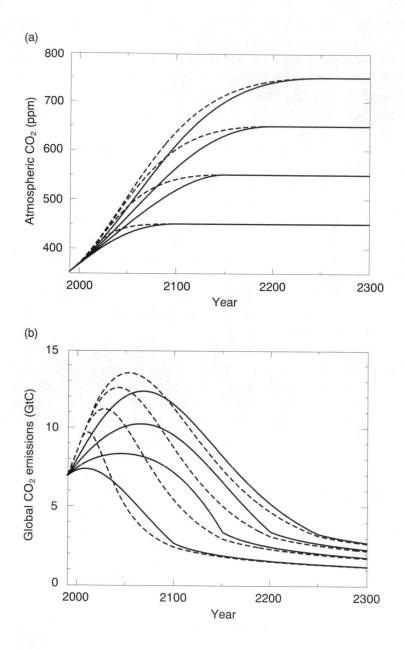

Figure 1.6. (a) Pathways for stabilizing atmospheric CO_2 concentrations at various levels. Dashed curves show alternative pathways to the same stabilization levels. (b) CO_2 emission pathways corresponding to concentration stabilization curves shown in (a). Source: Schimel *et al.*, 1996.

As a guide to this process, the IPCC has published a number of scenarios in which CO_2 concentrations stabilize at various levels via alternative concentration pathways (*Figure 1.6a*). Not only the concentration target, but also the choice of pathway to a given target, could have considerable consequences for the resulting rate of climate change and economic costs (Wigley *et al.*, 1995). *Figure 1.6b* shows the emission rates that would lead to each of the stabilization pathways shown in *Figure 1.6a*. All stabilization targets imply reductions from the IS92a "business as usual" trajectory within 50 years, but depending on the target and path that are chosen, emission rates could be allowed to more than double before beginning to decline in absolute terms. In addition, while all targets require emissions to eventually drop below the current global rate of 6 GtC/year after their initial period of growth, the timing of those reductions varies widely. Stabilization at 750 parts per million (ppm) may not require current emission rates to be reached for over 150 years, while stabilizing at 450 ppm would require reductions to current levels within about 40 years.

Using projections of future population size (see Chapter 2), the stabilization pathways can be translated into required global average per capita emission limits. Whether population follows a high- or low-growth path makes a significant difference to the results. For example, to stabilize CO_2 concentrations at 550 ppm, a high-growth path would require per capita emissions to fall to less than 0.1 tC per year by the end of the 21st century. In contrast, the low-growth path would allow per capita emissions to remain above 0.4 tC per year (Engelman, 1998). These differences translate into significant differences in the required decarbonization of the global economy (Bongaarts *et al.*, 1997).

Given the wide range of emission pathways that are consistent with stabilizing CO_2 concentration, the emission limits specified in the Kyoto Protocol are consistent with long-run concentrations in the 450–750 ppm range. However, analysis of possible MDC reduction pathways over this century demonstrates that stabilizing CO_2 concentrations will not be possible through MDC emission reductions alone (Wigley *et al.*, 1997); eventual LDC participation will be essential to meeting the goal of the FCCC.

1.7 Conclusion

The greenhouse effect provides the basis for current understanding of the broad outlines of the climate history of the Earth. The Earth's natural greenhouse effect is being enhanced by buildup in the atmosphere of GHG concentrations due to human activity. Continued emissions of GHGs – CO_2, CH_4, and N_2O principal among them – are expected to lead to a changing climate over the coming decades and centuries. How much and how fast climate will change remains uncertain;

likewise, regional details cannot be projected with confidence. Nonetheless, modeling studies and analogies with past climate provide a picture of potentially severe consequences: rising sea levels, more frequent and intense heat waves, changes in patterns of precipitation, droughts and floods, possible changes in storm frequency and intensity, and a small but real possibility of surprise events verging on the catastrophic.

How severe climate change turns out to be will depend on the path of future GHG emissions, as well as on the sensitivity of the climate system to increased radiative forcing. Current and historical emissions have been dominated by MDCs; the balance of future emissions is likely to tilt increasingly toward LDCs. Uncertainties in population and economic growth, as well as in the kind and amount of energy used, produce a large range of possible future emission paths.

The political response to the climate change issue is also likely to be a decisive factor. Although the potential of anthropogenic GHG emissions to alter the climate has long been recognized, it is only over the past 20 years that serious attention has been focused on the problem. Since the late 1980s, the issue has steadily grown in prominence, a progression that has been marked most recently by the signing of the Kyoto Protocol. Should the Protocol be ratified, the agreement it contains to limit MDC emissions would be a first step toward achieving stabilization of atmospheric concentrations of GHGs as called for in the FCCC. However, additional steps, including eventual participation by LDCs, will be necessary to achieve this goal.

Notes

[1] It is well known that the greenhouse analogy is inexact, since the glass in a greenhouse warms the interior mainly by inhibiting convective cooling while greenhouse gases inhibit radiative cooling. However, the comparison communicates the essential idea sufficiently well to be standard terminology.

[2] Observations of the relative humidity profile with changes in latitude and season show that relative humidity remains roughly constant with an increase in temperature (Manabe and Wetherald, 1967).

[3] For example, cumulative anthropogenic emissions are about twice the increase in the atmospheric content of CO_2, so the anthropogenic source is of sufficient magnitude. In addition, studies of the ratios of different isotopes of carbon in atmospheric CO_2 over time show patterns that are well explained by an anthropogenic source of rising CO_2 levels but are inconsistent with a natural source. Also, natural sources were evidently in balance for thousands of years prior to the Industrial Revolution, as the atmospheric CO_2 concentration over that period varied by only ± 10 ppm around its mean value of 280 ppm. Despite this overwhelming evidence, the argument for a natural source still appears from time to time in political debates on climate change.

[4] The interpretation of GWPs in terms of climate change, as opposed to changes in radiative forcing, is another matter. Whether GWPs are a reasonable index of climate

response to emissions of different GHGs is an open question (O'Neill, in press; Smith and Wigley, in press).

[5] The steady rise in CO_2 concentration shown in *Figure 1.1* appears to be inconsistent with the uneven rise in global average temperature shown in *Figure 1.2*. When trends in sulfate aerosol concentrations are considered in addition to trends in CO_2 concentrations (see *Figure 1.3*), the correspondence between temperature and atmospheric composition is closer. Natural climate variability has likely also played a role.

[6] Most of the stratospheric cooling is due to ozone depletion; however, observed temperature declines are steeper than would be expected from ozone losses alone (Shine *et al.*, 1995).

[7] Carbon equivalent emissions are calculated using 100-year global warming potentials (GWPs) as given in *Table 1.1*. For example, carbon equivalent emissions of gas X are calculated as $E_{X,ceq} = E_X \times GWP_X$.

[8] Throughout, we follow the conventional practice of referring to countries as more developed countries (MDCs) or less developed countries (LDCs) on the basis of their current status. Yet over the long time frame relevant to climate change, some countries currently defined as LDCs will eventually attain levels of development that would qualify them as MDCs today. Some LDCs, by sustaining rapid economic growth, will likely become MDCs by future standards as well. Thus, the terms "MDC" and "LDC" should not obscure the fact that almost all countries are likely to be better off in the future than they are today, nor is it implied that the membership of each group will remain constant. However, the historical record holds out little hope that poor countries, by growing faster than rich countries, will converge from below on a uniformly high level of per capita income (Pritchett, 1997).

[9] See Banuri *et al.* (1996) for a discussion of the controversy over this point generated by a 1990 report by the World Resources Institute.

[10] Wigley (1999) projects a somewhat higher range of temperature increases (1.3–4.0°C) by the year 2100 based on preliminary versions of a new set of IPCC emission scenarios. The difference is due principally to a reduction in projected sulfur emissions.

[11] Evapotranspiration describes the two main processes by which land surfaces lose water to the atmosphere: evaporation and transpiration by plants.

[12] See Weiner (1990) and Weart (1997) for interesting accounts of the early history of scientific research on the greenhouse effect.

Chapter 2

The Human Population

Dramatic changes in demographic patterns have taken place over the past several decades, particularly in less developed countries (LDCs). Mortality and, subsequently, fertility have fallen at rates much faster than rates of decline experienced in the industrialized countries. Unlike the more developed countries (MDCs), demographic transition from high to low fertility is not yet complete in many LDCs. In this chapter we summarize the current demographic situation and discuss factors behind the demographic transition in LDCs. This discussion provides background for the exploration of population–development–environment interactions taken up in Chapter 3 and in Part II of the book.

In addition, we discuss projected trends in population growth and aging over the 21st century based on the most recent projections of the International Institute for Applied Systems Analysis (IIASA). There appear to be three virtual certainties: world population will rise significantly from its present level of around 6 billion, although it may start to decline during the second half of the next century; its distribution will continue to tilt from MDCs to LDCs; and it will continue to age. Probabilistic projections indicate that global population is unlikely to double from its present size; on the other hand, a severalfold increase in the size of the population above age 60 is virtually assured. The 60+ age group, measured as a proportion of total population, will almost certainly double and will likely triple during the course of this century.

The most important source of uncertainty in population projections is the future path of fertility, although the impacts of alternative trends in mortality and migration are significant as well. We discuss the assumptions for each of these three demographic components, which underlie IIASA's projection results. Variations in the paths of these variables can give rise to large differences in future population size and age structure. For example, in 2100, the 95% confidence interval for global population is 5.7 to 17.3 billion. Although this level of uncertainty might be

39

considered high, relative to other factors driving global environmental change, population is one of the more robust (i.e., insensitive to changes in assumptions) variables to be considered. At the same time, as discussed in Chapter 4, this range has the potential to significantly change the outlook for future greenhouse gas (GHG) emissions and climate change.

2.1   Demographic Trends: A Global Summary

At the dawn of the agricultural revolution, 8,000 years ago, total world population was about 250,000 (Cook, 1962). It was not until after 1800 that the population reached 1 billion (roughly the current population of Europe and North America combined). It took more than a century, until 1930, to add the second billion. It took only 30 years, until 1960, to add the third billion. The fourth billion was added between 1960 and 1975, and the five-billion mark was passed in 1987. The sixth billion will be added by late 1999.

In fact, despite the apparent acceleration, both the annual growth rate of world population and the number of persons added each year have passed their peaks and are expected to continue falling. The growth rate peaked at 2.1% per year in the late 1960s and has since fallen to 1.5% (see *Table 2.1*), and the annual absolute increment of population peaked at about 87 million per year in the late 1980s and is now about 81 million. This does not mean that little additional population growth is to be expected; most mid-range population projections foresee future population rising to 10–12 billion by the end of the 21st century. The statistics are, however, indicative of the tremendous changes that have taken place over the past four decades.

As shown in *Table 2.1*, the total fertility rate (TFR) declined modestly in most parts of the world during the 1950–1955 to 1970–1975 period, then declined over the following 20 years with a rapidity that was unimaginable in the 1960s.[1] This second period of decline was especially pronounced in Asia, where the TFR fell by more than two children per woman (a statistic that is, however, heavily influenced by a dramatic fertility decline in China during the 1970s). One exception has been Africa, where fertility rates remained well above six children per woman through the late 1980s; since then, the beginnings of a fertility decline have become apparent. Meanwhile, regions like Europe and North America that had already achieved very low fertility by 1970–1975 saw these rates persist or fall even further.

During the 1950s and 1960s, reductions in mortality resulting from the spread of modern sanitation and medicine were more significant than fertility declines. During the period 1950–1955 (the first period for which estimates are available), life expectancy was lowest in Africa (38 years) and Asia (41 years), but had already improved significantly in Latin America (51 years).[2]   Over the subsequent 20

Table 2.1. World demographic trends since 1950.

	Total population size (millions)			Growth rate (%/yr)			Life expectancy (yrs), both sexes combined			Total fertility rate		
	1950	1970	1995	1950–55	1970–75	1990–95	1950–55	1970–75	1990–95	1950–55	1970–75	1990–95
World	2,524	3,702	5,687	1.78	1.95	1.48	46.5	57.9	64.3	5.00	4.48	2.96
MDCs	813	1,008	1,171	1.21	0.79	0.40	66.5	71.2	74.2	2.77	2.11	1.68
LDCs	1,711	2,694	4,516	2.05	2.37	1.77	40.9	54.7	62.1	6.17	5.42	3.30
Africa	224	364	719	2.23	2.56	2.68	37.8	46.0	51.8	6.64	6.57	5.71
Asia	1,402	2,147	3,438	1.91	2.27	1.53	41.3	56.3	64.5	5.90	5.09	2.84
Europe	547	656	728	1.00	0.60	0.16	66.1	70.8	72.7	2.56	2.14	1.57
Latin America and Caribbean	166	284	477	2.65	2.43	1.70	51.4	61.1	68.5	5.88	5.01	2.93
North America	172	232	297	1.70	1.01	1.31	69.0	71.5	76.2	3.47	2.01	2.02
Oceania	13	19	28	2.21	2.09	1.37	60.9	66.6	72.9	3.84	3.21	2.51

Source: United Nations, 1997.

years life expectancy increased impressively in all parts of the world. In Asia, by far the most populous continent of the world, it increased by 15 years over this short period. In Africa, it improved by 8 years, although this increase was below the world average. Improvements continued over the next 20 years to 1990–1995 but at a somewhat slower speed, with Asia, Latin America, and Africa all seeing substantial improvements. Very recently, some African countries seriously affected by the human immunodeficiency virus (HIV)/acquired immunodeficiency syndrome (AIDS) epidemic have seen reversals of this trend.

The rapid increases in life expectancies since the 1950s have been mostly due to falling infant and child mortality rates, with more children surviving to adulthood. Combined with fertility rates that remained high (or even increased somewhat due to better maternal health), the decline in infant and child mortality led to soaring population growth rates in the 1950s and 1960s. As shown in the table, growth rates in 1950–1955 were highest in Latin America because high fertility in the region was associated with already-lower infant and child mortality. By 1970–1975, however, the population growth rate in Africa surpassed that in Latin America as Africa's mortality rates fell and fertility rates hovered at about 6.6 children on average per woman. Because fertility rates have remained relatively high in this region, population growth accelerated to 2.7% in 1990–1995. Were this growth rate to remain constant, population would double in 26 years.

These trends in fertility and mortality resulted in different patterns of population growth in different parts of the world. In fact, the dominant feature of the global demographic landscape has been the contrast between the well-off populations of Europe, North America, and Japan and the poorer populations of Asia, Africa, the Middle East, and Latin America. The population of the industrialized MDCs is relatively small (about 1.2 billion in 1995) and is expanding very slowly (0.4% per year) following a 44% increase since 1950 (see *Table 2.1*). That of the LDCs is large (about 4.5 billion in 1995) and expanding rapidly (1.8% per year) after increasing by a factor of 2.6 since 1950. As a consequence, the share of today's industrialized countries in the world population decreased from 32% in 1950 to 21% in 1995 and is likely to decrease much more in the future. In addition, despite the rapid changes in most LDCs, inhabitants of the MDCs on average live significantly longer (life expectancy at birth for both sexes combined is about 74 years, versus 62 in LDCs) and have fewer children (TFR is 1.7, versus 3.3 in LDCs).

The widely varying historical experiences of the different regions of the world have left a strong imprint on the age structures of their populations. Age structures are often depicted as "age pyramids" showing the number of men and women in five-year age groups. *Figure 2.1* shows age pyramids for the two most extreme cases, sub-Saharan Africa and Western Europe. In Africa the pyramid is typical of a rapidly growing population, showing larger and larger cohorts at the bottom

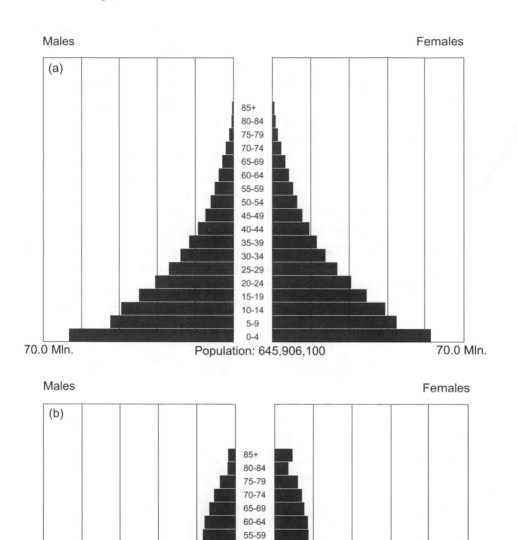

Males   Females

(a)

85+
80-84
75-79
70-74
65-69
60-64
55-59
50-54
45-49
40-44
35-39
30-34
25-29
20-24
15-19
10-14
5-9
0-4

70.0 Mln.   Population: 645,906,100   70.0 Mln.

Males   Females

(b)

85+
80-84
75-79
70-74
65-69
60-64
55-59
50-54
45-49
40-44
35-39
30-34
25-29
20-24
15-19
10-14
5-9
0-4

70.0 Mln.   Population: 457,320,000   70.0 Mln.

Figure 2.1. Age pyramids of (a) sub-Saharan Africa and (b) Western Europe in 2000. Source: United Nations, 1997.

(i.e., in the young age groups). In Western Europe the pattern is completely different; the population is distributed more equally across age groups.

The narrowing of population pyramids at the bottom (from low fertility) and fattening at the top (due to extended longevity) is illustrative of the change in population age structure that has come to be called "population aging." The two components are referred to as aging "from the bottom" and "from the top." Population aging is an important social phenomenon, especially in relation to the uncertain future of pension and health care systems and implications for labor markets and the work force (Kinsella and Gist, 1995). Aging will continue in MDCs and has already started in LDCs. Just as the speed of mortality improvements accentuated the implications of demographic transition for population growth rates, the speed of LDC fertility decline will accentuate the aging phenomenon.

Another major structural change in world population is the move from a predominantly rural society (at the global level) to an urban one (see *Box 2.1*). Especially in the poorest LDCs, where economic and social policies have been geared toward agriculture and rural development, there will need to be a rebalancing of priorities.

2.2 Fertility Decline in Less Developed Countries

The outstanding demographic event of the last quarter of the 20th century was the rapid decline in fertility around the world, particularly in LDCs. To understand this change and to provide a framework for understanding the assumptions about future fertility that are critical to population projections, this section reviews current thinking on the forces driving demographic changes.

2.2.1 Demographic transition

The demographic transition began in MDCs in the late 18th century and spread to LDCs in the last half of the 20th century (Notestein, 1945; Davis, 1954, 1991; Coale, 1973). The conventional theory of demographic transition predicts that as living standards rise and health conditions improve, mortality rates decline and then, somewhat later, fertility rates decline. Demographic transition theory has evolved as a generalization of the sequence of events observed in what are now MDCs. In these countries, mortality rates declined comparatively gradually beginning in the late 1700s and then more rapidly in the late 1800s; fertility rates declined, as well, after a lag of 75–100 years. Different societies experienced transition in different ways, and today various regions of the world are following distinctive paths (Tabah, 1989). Nonetheless, the broad result has been a gradual transition from a small, slowly growing population with high mortality and high

---

Box 2.1. Urbanization.

According to the United Nations (1998a), in the first decade of the 21st century, for the first time in history more people will live in cities than in rural areas. The urban proportion of the world population has increased from 30% in 1950 to 46% in 1996, and is forecast to reach 61% by 2030. In absolute terms rural population is expected to grow until 2020 and then begin a slow decline.

The pace of urbanization differs across regions. In MDCs, the proportion of the population living in urban areas increased from 55% to 75% between 1950 and 1995. The increase was sharper for developing regions, where the proportion more than doubled, rising from 18% in 1950 to 38% in 1995. As a result the LDC share of the world's urban population grew from 40% to 66% over this period.

Within LDCs, the level of urbanization is positively correlated, and the pace of urbanization is inversely correlated, with per capita income. In Latin America, which has the highest per capita income of any developing region, the transition to an urbanized society is far advanced. In fact, Latin America, where 73% of the population is urban, is indistinguishable from the MDCs in this regard. In Africa, the poorest region, the process is still in its early stages. The urban share of the African population increased from 17% in 1975 to 35% in 1995.

A striking feature of the urbanization process has been that primate cities (i.e., cities at the top of national city-size distribution) have grown more rapidly than secondary ones. Rapid urbanization has thus resulted in the emergence of a class of "mega-cities" (more than 10 million inhabitants). In 1950, only 2% of urban dwellers lived in cities of over 10 million inhabitants; in 1995, in contrast, the figure was 7%. The distribution of large cities between MDCs and LDCs has also changed markedly. In 1950, 11 of the 15 largest cities in the world were in MDCs; in 1995, only 4 (Tokyo, New York, Los Angeles, and Osaka) were in MDCs.

What drives urbanization? Economic theory interprets urbanization as a response to efficiencies and cost advantages that arise when economic activities are undertaken in close proximity to each other. Eventually, diseconomies of scale and urban disamenities – pollution, congestion, etc. – should cause the growth of mega-cities to slow. This slackening of growth might not take the form of people moving back to rural settings; rather it might reflect a situation in which smaller cities grow more rapidly than mega-cities. Some evidence exists that population "deconcentration" of this type, which was observed in the USA and Europe in the 1970s and 1980s, commences at an income level of roughly US$5,000 (MacKellar and Vining, 1995). The evidence is weak, however, and even if it is true, most LDCs have far to go in the development process before they reach the income level at which "deconcentration" could be expected.

fertility to a large, slowly growing population with low mortality and low fertility. During the transition itself, population growth accelerates because the decline in death rates precedes the decline in birth rates.

On the theoretical level there are two different ways to explain demographic transition. In one view, the fertility decline is a direct response to the mortality decline. This so-called homeostasis argument claims that societies tend to find an equilibrium between births and deaths. When death rates decline due to progress in medicine and better living conditions, the equilibrium is disturbed and the population grows unless birth rates decline in response to the new mortality conditions. The fact that fertility tends to decline many years after a decline in mortality may be explained by a perception lag. The alternative to the homeostasis argument assumes that modernization of society acts as a driving force of both declining mortality and declining fertility. Fertility decline lags behind mortality decline, according to this view, because fertility is embedded in the system of cultural norms more strongly than mortality and therefore changes more slowly. The historical record of Europe – where fertility sometimes declined simultaneously with mortality and population growth was generally much lower than in today's high fertility countries – tends to support the second explanation. But the two arguments are not necessarily mutually exclusive.

Figure 2.2 illustrates the example of demographic transition in Mauritius, a developing country with good records for birth and death rates going back more than a century. Until about 1945, annual birth and death rates fluctuated widely due to epidemics, the changing severity of endemic diseases (notably malaria), and changing weather conditions. Whenever birth rates are consistently above death rates, as was the case in Mauritius during the late 19th century, the population grows. After 1945, death rates in Mauritius declined precipitously due to malaria eradication and the introduction of modern medical technology. Birth rates, on the other hand, remained high or even increased somewhat due to the better health status of women (a typical phenomenon in the early phase of demographic transition). By 1950 this had resulted in a population growth rate of more than 3% per year, one of the highest in the world at that time. Birth rates subsequently declined, with the bulk of the transition occurring during the late 1960s and early 1970s when the TFR declined from more than six children per woman to less than three within only seven years, probably the world's most rapid national fertility decline. This decline occurred on a strictly voluntary basis and resulted from the combination of high female educational status and well-implemented family planning programs (Lutz, 1994). Because of the still very young age structure of the Mauritian population, current birth rates are still higher than death rates and the population is growing by about 1% per year despite fertility at around replacement level.[3]

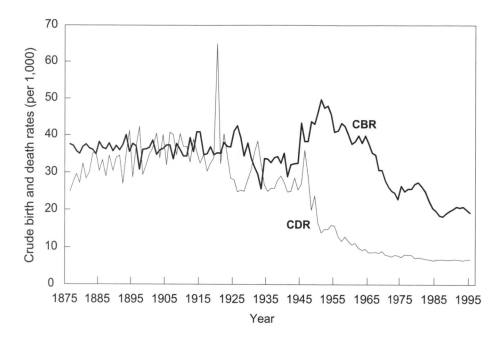

Figure 2.2. Crude birth rate (CBR) and crude death rate (CDR) in Mauritius, 1876–1996. Source: Mauritius Central Statistical Office.

Empirically observed trends in all parts of the world have overwhelmingly confirmed the relevance of the concept of demographic transition to LDCs. With the exception of pockets where religious or cultural beliefs are strongly pronatalist, fertility decline is well advanced in all regions except sub-Saharan Africa, and even in that region many signs of a fertility transition can be perceived. In Southeast Asia and many countries in Latin America, fertility rates are on a par with rates seen in MDCs only several decades ago, and in several countries, such as China, Taiwan, and Korea, fertility is at sub-replacement levels.

The biggest difference between the demographic transition processes in what are now the MDCs and LDCs has been the speed of mortality decline. Mortality decline in Europe, North America, and Japan came about over the course of two centuries as a result of reduced variability in the food supply, better housing, improved sanitation, and, finally, progress in preventive and curative medicine. Mortality decline in LDCs, in contrast, occurred very quickly after World War II as a result of the application of Western medical and public health technology to infectious, parasitic, and diarrheal diseases. Life expectancy in Europe rose gradually from about 35 years in 1800 to about 50 in 1900, 66.5 at the end of World War II, and 74.4 in 1995. In LDCs, it shot up from 40.9 at the end of World War II to 62.1 in 1995. The increase that took MDCs about one and a half centuries to achieve came to pass in LDCs in less than half a century. As a result of the speed of the mortality decline, populations in LDCs are growing three times faster today than did the populations of the present-day MDCs at the comparable stage of their own demographic transition.[4]

2.2.2   Fertility decline: Determinants and preconditions

Studies of the factors influencing changes in fertility must begin with the proximate determinants of fertility: (1) age at first marriage (or beginning of sexual activity); (2) prevalence and effectiveness of contraception; (3) prevalence of induced abortion; and (4) duration of postpartum infecundibility, especially due to breast-feeding (Bongaarts and Potter, 1983).[5]   Fertility decline must occur through changes in one or more of the four proximate determinants.

The adoption of contraception has been the principal source of fertility decline in LDCs. However, how couples adopt contraceptive practices is a function of many influences. The spread of contraceptive practice is a diffusion process consisting of stages of awareness, information, evaluation, trial, and adoption. All of these stages consist of actions undertaken in social networks, leading to path-dependence and the persistence of heterogeneity between subpopulations (Kohler, 1997). Coale (1973) lists three "preconditions" required for fertility decline. First, fertility must be regarded as being within the realm of conscious choice. Often, this marks a fundamental change in the way individuals view their lives and their families (Lockwood, 1997; van de Walle, 1992); for example, people may change from having a fatalistic attitude toward fertility to making procreation an object of their life-course planning. Yet, in most demographic transitions, fertility regulation was already practiced during the pretransition phase, albeit more for spacing than for limiting the final number of children (Mason, 1997). Therefore, second, there must be objective advantages to lower fertility. And third, acceptable means of fertility reduction must be at hand. These three preconditions for a lasting fertility decline suggest three parallel strategies to foster the transition from high to low fertility:

- Emphasize universal basic education to bring fertility increasingly into the realm of conscious choice. Mass media may also exert an important influence.
- Pursue changes in socioeconomic variables, mostly economic costs and benefits associated with child labor, female participation in the modern-sector labor force, support in old age, etc. Changes in the "value" of children have an impact on couples' desired family size.
- Invest in reproductive health and the availability of family planning services, including maternal and child health programs that reduce infant and child mortality. Help couples match their desired and actual number of children by focusing on the unmet need for family planning (see *Box 2.2*).

This framework suggests that if two of the three preconditions are already met, the introduction of the third may trigger a rapid fertility decline. In the case of the rapid Mauritian fertility decline, the young female population was already literate and large families were increasingly perceived as an economic burden. The strong voluntary family planning campaign, which strengthened the negative perception of high fertility and provided efficient family planning services that were even supported by the influential Roman Catholic Church, then triggered the precipitous fertility decline. In other countries, in contrast, huge investments in family planning were virtually without effect because one or both of the other two preconditions were not met.

This view of the three necessary preconditions offers a different perspective on the fundamental "chicken and egg" ambiguity regarding causation: does provision of family planning through national programs lead to lower fertility via increased adoption of contraception, or do parents' declining fertility desires, translated into demand for contraception, induce a supply response in the form of a national family planning program? The difference between the positions was well captured by the rallying cries of LDC and MDC delegates to the 1974 Population Conference held in Bucharest: "Development is the best contraceptive," on the one hand, and "Contraception is the best contraceptive," on the other. A new position, which emerged forcefully at the 1994 International Conference on Population and Development in Cairo, might be expressed as "Empowerment of women is the best contraceptive." Generally speaking, the point of view emphasizing the role of national family planning programs was supported by members of the international family planning community. The point of view emphasizing economic development and shifting economic costs and benefits of childbearing was supported by neoclassical economists (as well as LDC politicians striving for rapid economic growth). The point of view emphasizing reproductive rights and the status of women was supported by women's advocates. Each perspective has insights to offer. All three together are overwhelmingly likely to make a difference.

National Family Planning Programs

The proximate cause of fertility decline is not in dispute: fertility decline in the Third World has occurred as rising proportions of married women have adopted contraception (Weinberger, 1994).[6] The statistical association between various indices of what the international public health community calls "family planning program effort" and the prevalence of contraception is indisputable at the national level. From comparing fertility trends in broadly similar countries, such as Mexico and Colombia, and Pakistan and Bangladesh, Cleland (1996) was led to conclude that government attitudes toward family planning were a crucial variable. However, experience over several decades has shown that family planning programs have an

---

Box 2.2. Unmet need for family planning.

The most common strategy for addressing the unmet need for family planning –
defined as the number of women at risk of unwanted pregnancy who are not
practicing modern contraception – is subsidized provision of voluntary family
planning services. Family planning programs seek to solve problems that generally involve insufficient information about or access to contraceptive services
(Chomitz and Birdsall, 1991). Programs may also go beyond provision of contraceptive methods and information by addressing social obstacles such as disap-
proval from religious communities or husbands.

    Family planning programs have had a significant impact on fertility rates over
the past three decades. Bongaarts (1990) estimated that LDC fertility would have
fallen from an average of about 6 births per woman in the 1960s to 5.4 in the
1980s in the absence of such programs; with them, TFR actually fell to about 4.2.
As a result, the LDC population was 400 million less in 1990 than it would have
been without family planning programs. Unmet need is estimated at about 100
million women in LDCs, which leads to the often-cited estimate that one in every
four LDC births (excluding China) is unwanted (Bongaarts, 1990). Eliminating
all unwanted births would reduce LDC population in 2100 by about 2 billion
(Bongaarts, 1994a).

    However, Pritchett (1994a) argues that differences in fertility rates across
countries are largely attributable to differences in desired fertility, not to differences in family planning program efforts. Future declines in fertility would therefore have to come largely from reducing desired family size rather than reducing
unmet need, an assertion that has touched off a lively debate (Bongaarts, 1994a;
Knowles *et al.*, 1994; Pritchett, 1994b). Although the relative influence of desired
fertility and unmet need is an important matter, as we discuss in the text, no single
factor contributing to fertility decline is likely to be effective in the absence of the
others.

---

impact only when desired fertility has already declined due to socioeconomic development (Bongaarts *et al.*, 1990) and that such programs work best when integrated
with other government policies in areas such as health, education, and rural development (Concepcion, 1996). The mere supply of contraceptive technology backed
up by government propaganda is no guarantee of change. Even in cases where
fertility decline has occurred despite any evident change in economic variables or
cultural values, it can be argued that decline in infant and child mortality had produced latent demand for lower fertility (Cleland, 1993; Cleland and Wilson, 1987).
As a general proposition, then, national family planning programs are at best a

necessary, but never a sufficient, condition for fertility decline. On the other hand, assuming that fertility is already within the calculus of conscious choice and the socioeconomic preconditions for lower fertility are present, well-designed family planning programs can greatly accelerate fertility decline.

Economic Costs and Benefits

The neoclassical economic model of fertility explains childbearing behavior as the outcome of a utility-maximization decision made by parents in response to economic costs and benefits (Becker, 1981; Becker and Barro, 1988; Becker *et al.*, 1991). The economic benefits from children are mainly of two types: (1) current income flows, which can consist of the imputed value of home labor, extramural earnings turned over to parents by children still resident in the household, or remittances sent by children who have left the household (Clay and Vander Haar, 1993) and (2) future old-age security. Economic costs relate mostly to the costs of education and the opportunity costs of time expended in childbearing and child care. In the Philippines, for example, women's contraceptive use is lower in communities where children make significant economic contributions but higher in communities where women have favorable labor market opportunities (De Graff *et al.*, 1997).

According to the neoclassical model, parents' well-being is considered to be a function of "child quantity" (i.e., number of children), "child quality" (i.e., educational endowment, health, etc.), and the level of consumption of a third, composite good. The price of child quantity involves fixed costs, such as time expended in childbearing; the price of child quality involves costs related to education and other investments in human capital. The availability of natural resources and environmental assets can be dealt with by adding these to the utility function, on the assumption that parents make a four-way trade-off between number and quality of children, material consumption, and state of the environment.[7] Altruism toward future generations can be added to the model by including well-being of the children as an additional argument in the parents' utility function (Nerlove *et al.*, 1987b).

Fertility decline, according to the neoclassical model, results from a combination of increasing costs of child quantity (for example, an increase in the formal-sector labor-force participation rate of women, which raises the opportunity costs of time expended in child care) and declining costs of child quality (for example, increasing availability of educational opportunities). All else being equal, gains in income per capita associated with economic development will lead to increases in fertility just as they lead to increases in the consumption of material goods. All else is not equal, however, since development is accompanied by shifts in the relative prices of child quantity and quality. To these price shifts must be added direct

reductions in the economic rate of return on children as assets: modernization reduces the importance of child labor, and the opening of avenues for financial saving
and the provision of public social security schemes reduce the value of children as
pension and insurance assets. Shifts in relative prices combined with declines in the
asset value of children cause households to shift consumption from child quantity
to child quality.

Status and Empowerment of Women

It is consistently observed in a large number of surveys [e.g., the World Fertility Survey (WFS) and the Demographic and Health Survey (DHS)] that women's
desired number of children is lower than their husbands'. Women bear a disproportionate share of the costs of an additional birth (health risks, work associated with
child rearing, etc.) while husbands enjoy a disproportionate share of the associated
social prestige. Higher social status of women within the family and in society can
be expected to empower women in pursuing their own, lower desired family size
rather than that of their husbands. In addition, increasing women's social status is
likely to involve attitudinal shifts and changes in women's ideals and aspirations,
including reduction in desired fertility. Hence, higher status of women will have a
doubled effect promoting lower fertility.

Synthesizing the Views

What policy-relevant synthesis can be distilled from the above discussion? All
research suggests that fertility decline is complex, regionally differentiated, and
context dependent. Generalizations are risky even within regions, let alone worldwide. But if any generalization can be made, it is that female basic education is of
importance with respect to all three factors: status and empowerment of women,
economic costs and benefits of fertility, and acceptance of the services offered by
national family planning programs.

The strongest social correlate of fertility (far stronger than income, occupation,
labor-force participation, and the like) has long been known to be female education
(Cochrane, 1979; Cleland and Wilson, 1987; Cleland and Rodriguez, 1988).[8]
Even when controlling for socioeconomic status, the independent effect of education exceeds that of any other factor (United Nations, 1987, 1993).[9] One part
of this impact is an accounting effect: if girls are attending secondary school, age
at first marriage, one of the principal proximate determinants of fertility, is higher.
But more important is the effect of education on the values and tastes that provide the context for fertility decisions. The strongest intermediate link between
female educational attainment and lower fertility is the greater likelihood that married women who were schooled as girls will use contraception. While there are

increasing returns (in terms of lowered fertility) as the number of years of schooling rises, women who have had even a few years of primary schooling are observed to be more likely to practice contraception than those who have had none.

Part of the effect of girls' education is economic. It increases the cost of children by making girls unavailable for domestic chores; educated mothers are also more likely to earn income outside the home than uneducated ones, and hence face higher opportunity costs in child rearing. But there is more to education of girls than just shifting economic costs and benefits. Parents' decision to endow girls with even a modicum of education, as opposed to none at all, reflects a fundamental shift in the status of women. It sets in motion an irreversible shift in women's perceptions, ideals, and aspirations (Cleland and Wilson, 1987). Women who have been educated are better able to follow their own, usually lower, fertility desires rather than those of their husband or their extended family.[10] Closely related to the low status of women are poverty and insecurity, both of which raise the value of children as economic assets (in particular, as sources of old-age security) and impede the attitudinal change necessary for fertility decline (Thomas, 1991). Finally, provision of basic female education promotes a political culture and civil society in which national family planning programs are likely to be established and function effectively.

What is clear is that there is synergism between the various factors – national family planning programs combined with socioeconomic development and empowerment of women result in fertility decline. Investing in female basic education has positive effects on all three factors on top of its strong direct link to lower fertility. At the primary level, costs of providing basic education are extremely low. For policymakers who are attempting to accelerate demographic transition by encouraging fertility decline, it is the best candidate for priority action.

2.3 Population Projections

Virtually all demographers believe that population projections should be performed by using the so-called cohort-component model, which projects the population by age group. This method was introduced by Whelpton (1936), formalized by Leslie (1945), and first applied to project the world population by Notestein (1945). Apart from computational changes, the original method has not changed except for generalization to so-called multistate projections, which simultaneously project several subpopulations such as population by region, educational group, etc. Cohorts are groups of men and women born during the same time interval and representing a specific age group at any later point in time. Cohort size can only change through mortality (and, if also defined by region, through migration). The size of the youngest age group is determined by age-specific fertility rates applied to the corresponding female cohorts of childbearing age. Hence the projected size and age structure of the population at any point in the future depends exclusively on the size and age structure in the starting year combined with assumed age-specific fertility, mortality, and migration rates over the projection period. These three variables determine all future population trends.

2.3.1   Population momentum

Population growth is characterized by a momentum that arises from the persistence of a population's age structure. The youthful age structure of a high fertility population ensures that it will continue to grow for many years even if the TFR declines rapidly, since for a generation progressively larger cohorts of young women will be entering their reproductive years (see *Figure 2.1*). Age structure has this persistent effect because it takes 20 to 30 years for a cohort of newborn girls to reach their main reproductive ages; as a result, a decline in fertility rates does not show up as a decline in the number of potential mothers for several decades. Population momentum has two important implications. First, just as a braking vehicle only gradually comes to a halt due to momentum, population size reacts only very slowly to shifts in fertility parameters. For example, if LDC fertility rates were suddenly to decline to replacement levels, population in these countries would still continue to grow for several decades before leveling off. Under an instant replacement fertility scenario for the world, population would still grow to above 8 billion (United Nations, 1993). Second, just as inertia allows the motion of a braking (or accelerating) vehicle to be predicted with a relatively narrow margin of error, demographers can project population several decades into the future with a fair degree of confidence. This does not arise from any methodological sophistication or particularly impressive insight; it is mostly due to the fact that a large part of the population that will be alive 25 years from now has already been born. The remaining margin of uncertainty is only due to fertility, mortality, and migration over the coming years. Of these three, migration is the most volatile since it depends in part on political conditions; fertility and mortality tend to change more slowly over time.

2.3.2   Cumulation of errors and long-term projections

Compared with near-term projections, long-term projections are affected more significantly by uncertaintics in projected trends of fertility, mortality, and migration. Because of the nature of the compounding process, small errors in the early years of a projection (or small errors in the initial-year data on which the projection is based) translate into large inaccuracies in the long run.

Long-term projections are therefore more uncertain than those made for the next few decades, and extremely long-run population forecasts are little more than rational speculations. If reasonable confidence intervals are attached to these projections (as is done below), the range of possible outcomes fans out over time. Very small variations in the path of fertility over the next few decades are of particular concern, because they will eventually exert a disproportionate influence on the size of the population. Uncertainties in mortality are not as important, because high mortality in infancy and childhood has been substantially reduced in all regions so that variations in mortality rates do not generally (except for noninfectious disease mortality) make much difference in the numbers of women who survive to bear children.[11]

2.4 The IIASA 1996 Population Projections: Methodology and Assumptions

In this section, we summarize the assumptions and results of the population outlook published by IIASA (Lutz, 1996). These projections differ from others, such as United Nations (UN) projections, in three respects: (1) there is extensive discussion of the basis for assumed fertility, mortality, and migration trends; (2) alternative paths of future mortality and migration, in addition to fertility, are incorporated into the projections; and (3) probabilities are attached to alternative population paths.

2.4.1 Methodology

It has become the standard practice of most national and international agencies that prepare population projections to provide the user with a "medium variant" resulting from a set of assumptions considered "most likely," plus "low" and "high" variants based on different assumptions and resulting in lower and higher total population sizes, respectively. Although such low and high variants are sometimes misunderstood as providing confidence intervals in a probabilistic sense, they typically are not based on any systematic variation of parameters, but simply represent alternative sets of fertility assumptions. For example, the biannual projections prepared by the United Nations Population Division (United Nations, 1997) follow this approach. For every country in the world the UN presents three variants based on three different sets of fertility assumptions. The UN does not publish projections based on alternative mortality or migration assumptions, nor are users provided with a substantive justification of the assumptions made in the projections. In addition to the lack of a clear probabilistic interpretation of the high and low variants, at the global level these high and low variants implicitly make the strong assumption that all countries simultaneously follow either high or low fertility paths and thus do not allow for the possibility that diverging regional trends may offset each other. Finally, in its long-range medium variant the UN assumes that fertility in all regions of the world reaches and remains at replacement level, an assumption for which there is neither strong theoretical nor empirical support.

IIASA invited a number of experts on fertility, mortality, and migration to write chapters that summarized the existing demographic knowledge relevant for specifying the uncertainty range of future trends. These experts were asked to provide, for 13 world regions in the period 2030–2035, high and low estimates, designed to cover approximately 90% of the possible range, for fertility, mortality, and migration parameters. Central variant assumptions were taken as averages of the high–low range. Assumptions for 2030–2035 were extended to 2080–2085 as described below and were held constant thereafter.[12]

2.4.2   Assumptions on future fertility

In the long run, the level of fertility has the greatest impact on population growth because of its multiplier effect: additional children born today will have additional children in the future. Also, in the rapidly aging populations of Europe the effect of very low fertility – aging from the bottom – tends to be greater than the effect of increasing life expectancy – aging from the top.

Relying on the concept of demographic transition, population forecasters have conventionally assumed that every country will reach replacement level fertility sometime in the future. Traditionally this assumption of a convergence toward the replacement level has also been made for European populations that have already been far below that level for extended periods, implying increases in future fertility for these countries. This replacement level assumption is mechanistic, however, and is usually not supported by much substantive analysis.

Less Developed Countries

Assumptions about the future course of fertility are based on several sources. For the near future the intentions expressed by women in representative surveys can be taken as an important indication. In recent years such surveys – most important among them the WFS and the DHS – have been conducted in a large number of developing countries. In many countries these surveys provide more reliable information about actual fertility levels than the official birth registration system. But they also give information about family size preferences and intentions as well as the level of contraceptive use.

Table 2.2 lists this information for a selected number of countries. Sub-Saharan Africa has by far the highest fertility levels and the lowest levels of contraceptive prevalence. Within this region, countries with the lowest contraceptive prevalence (such as Mali and Uganda) have the highest fertility rates. But in all countries

Table 2.2. Actual and desired total fertility and contraceptive prevalence in selected countries with Demographic and Health Surveys (around 1990).

	TFR	Desired TFR	Contraceptive prevalence (%)
Sub-Saharan Africa			
Kenya	6.4	4.5	27
Mali	7.6	7.1	5
Senegal	6.6	5.6	11
Uganda	7.5	6.5	5
Zimbabwe	5.2	4.3	43
North Africa			
Egypt	4.4	2.8	38
Morocco	4.6	3.3	36
Tunisia	4.1	2.9	50
Asia			
Indonesia	2.9	2.4	48
Sri Lanka	2.6	2.2	62
Thailand	2.2	1.8	66
Latin America			
Bolivia	5.1	2.8	39
Brazil	3.3	2.2	69
Colombia	3.1	2.1	68
Mexico	4.0	2.9	58
Peru	4.0	2.3	54

Source: Westoff, 1996.

the desired fertility rates derived from the surveys are lower than the actual ones – in Kenya by almost two children. This, together with indications of increasing contraceptive prevalence, leads to the expectation of significant fertility declines in the near future even in sub-Saharan Africa.

In North Africa the fertility transition is already much further advanced, and more than one-third of the female population employs contraception. Desired fertility rates in this region indicate that further significant fertility declines are to be expected unless dramatic events, such as a rise in pronatalist traditional religious values, change the pattern. The three Asian countries considered here, Indonesia, Sri Lanka, and Thailand, already have very high levels of contraceptive prevalence and total fertility rates below 3.0, with further declines expected. Latin America has an intermediate position, comparable with that of North Africa.

For the longer-term future, general social and economic development, together with cultural and political factors, will determine the course of fertility decline. As discussed earlier in the chapter, the three driving forces that have been identified by research are national family planning programs, changing perception of the costs and benefits of childbearing, and the improving status of women.

Although our understanding of the causal mechanism at work is incomplete, the evidence favors further fertility decline in LDCs. Therefore, even high fertility variants in these regions generally call for fertility to be lower in the period 2030–2035 than it is today. One exception is China, where the high variant assumes that the country's one-child policy is relaxed to allow fertility more in line with desired fertility levels. High-variant fertility assumptions for other regions are that the ongoing fertility transition is retarded or stalled. That this can easily occur is illustrated by the important case of India, where fertility decline was arrested between the late 1970s and the early 1990s at a TFR of about 4. Low-variant assumptions imply that fertility decline in LDCs, as in MDCs, does not stop at a TFR of 2.1 but continues to decline, carrying countries into the range of sub-replacement fertility. The central assumptions, which represent the most likely case, result in fertility slightly above replacement level in 2030–2035 in most LDC regions – substantially so in sub-Saharan Africa.

Fertility assumptions were extended into the future by assuming that in 2080–2085, the TFR in all LDC regions will lie between 1.7 and 2.1, depending on the population density of the region. Although there is some empirical evidence that high population density tends to induce lower fertility after controlling for other factors (Lutz, 1997), this procedure is still somewhat ad hoc, although no more so than the common alternative, which is to assume that fertility falls (or rises, as the case may be) to replacement level and then remains fixed there.

More Developed Countries

Figure 2.3 plots the trends in TFRs for selected European countries since 1950. It shows the baby boom during the 1960s, which was a phenomenon common to all Western industrialized societies. The increase in births was due partly to a marriage boom caused by prosperous economic conditions, and partly to higher fertility within marriage, again related to economic prosperity. The 1970s were characterized by rapid fertility declines in almost all countries. The southern European countries lagged somewhat behind, but subsequently experienced an even steeper decline. Italy and Spain are currently experiencing the lowest fertility levels in Europe, with TFRs as low as 1.2 children per woman. In France and especially in Scandinavia fertility has never declined to such low levels. In Sweden, fertility showed a temporary increase to replacement level during the 1980s. This increase seems to have been associated with infrastructural arrangements (such as excellent day care) that facilitated the combination of child rearing and work outside the home. In the first half of the 1990s, however, fertility in Sweden came down again.

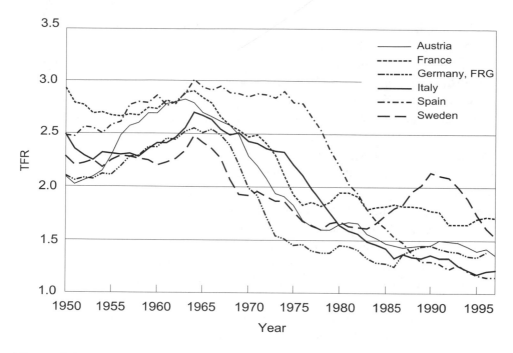

Figure 2.3. Total fertility rate (TFR) in selected European countries since 1950. Source: Council of Europe, 1997.

As the historical record in MDCs is mixed, the direction of future fertility trends is not clear. For their high variant, the IIASA projections assume a return to replacement level fertility (or slightly above) by 2030–2035.[13] This assumption, common to the central variants of other institutions discussed above, is theoretically supported by the homeostasis argument, which says that fertility levels are not so much the sum of individual behavior as one aspect of the evolution of a system where individual behavior is a function of the status of the system (see Vishnevsky, 1991). Under such a systems approach, the assumption of replacement fertility in the longer run makes sense.

It is difficult, however, to find many researchers who support this view. Too much evidence points toward low fertility. The return to replacement fertility has been criticized as an assumed magnetic force (Westoff, 1991) without empirical support, while there are many significant arguments supporting an assumption of further declining fertility levels. These can be grouped together under the term "individuation" (Bumpass, 1990), and they include the weakening of family ties both in terms of declining marriage rates and high divorce rates, the increasing independence and labor-market orientation of women, and the value change toward materialism and consumerism. Individuation, together with increasing demands and norms concerning the resources (attention, education, money, etc.) to be given to children, is likely to result in fewer couples with more than one or two children and an increasing number of childless couples.

A further argument for declining fertility is that surveys still show very high proportions of unplanned births, even in societies where contraception is universally available. For example, in the United States, about 30% of births are estimated to be unintended at the time of conception (Brown and Eisenberg, 1995; Montgomery, 1996; Henshaw, 1998). A third of these are unwanted (born to women who did not want any more children) and the rest are mistimed (born to women who did not want to become pregnant until a later time). Unplanned births make up a similar or greater proportion of total births in Canada and Japan (Alan Guttmacher Institute, 1999). If contraceptive methods without side effects were developed that required positive action in order to become pregnant instead of forethought to avoid pregnancy, the number of births would most likely decline further.

The bulk of evidence therefore points to continued low fertility or even further decline. How far it can decline is not clear. Golini (1998) recently suggested an absolute lower limit of about 0.7 to 0.8 children per woman based on the assumption that 20–30% of women remain childless and the rest have just one child. The IIASA projections set low values (which assume only a 5% probability of even lower levels) by 2030–2035 of 1.4 in North America and 1.3 in the other developed regions, with central variants obtained by averaging high and low assumptions.

2.4.3 Assumptions on future mortality

More so than for fertility, changes in the level of mortality are directly related to biomedical aspects of development. A behavioral component, however, enters through the choice of individual lifestyle (such as smoking), which, at least in low mortality countries, has a significant impact on individual survival probabilities.

The uncertainties about life expectancy in the future are quite different in today's high and low mortality countries. The latter have seen impressive increases in life expectancy, and it is no longer uncommon for individuals in MDCs to survive to ages that used to be considered a biological upper limit to the human life span. Assumptions about future improvements depend crucially on whether or not such a limit exists and will soon be reached. In regions that still have much lower life expectancy, this question is irrelevant and future mortality conditions will be determined by the effectiveness of local health services, the spread of both traditional (e.g., malaria) and new (e.g., AIDS) diseases, and the general standards of living and education. Given even modest social and economic development, and apart from the AIDS epidemic (see *Box 2.3*), it is likely that the gap between life expectancy in MDCs and LDCs will narrow further.

Box 2.3. AIDS and Africa.

The sheer number of deaths due to AIDS in Africa, and hence the magnitude of the epidemic on a human and social scale, is immense. By subjecting members of the most economically active and productive groups to a wasting and premature death, AIDS is imposing an enormous economic and social toll on the continent (Ainsworth and Over, 1994). However, the effects of AIDS on population growth rates are assumed to be modest in most countries. Bongaarts (1996) estimates that by 2000, the AIDS death rate in sub-Saharan Africa will approximately equal the HIV seroprevalence rate, implying that seroprevalence will plateau. In the worst case, this results in an increase of the crude death rate of 10.7 deaths per 1,000 members of the population; in the best case the figure is 1.7 deaths per 1,000 members of the population. Results are sensitive to assumptions regarding social response; for example, a simulation analysis by Cuddington et al. (1994) finds that in a hypothetical African population an increase in condom usage from 0% to 10% cuts steady-state AIDS prevalence by over one-third, from 31% to 19% of the population.

For a few African countries the situation seems to be more dramatic, however. Recent simulations for Botswana (Sanderson et al., forthcoming), which has an HIV prevalence rate estimated to be well above 30%, indicate that total population size may well decline due to AIDS. Moreover, the age structure will become severely distorted, and AIDS orphans, rising health expenditures, and a worsening health status of the labor force are likely to present major macroeconomic problems in addition to immense human suffering.

The ultimate impact of AIDS on the population of Africa as a whole will be moderate if, as assumed by Bongaarts (1996) and others, exceptionally high HIV prevalence rates exist only in Botswana and a few other countries with special conditions. If high prevalence rates turn out to be more widespread, AIDS could have a significant impact on African population dynamics.

Less Developed Countries

Figure 2.4 depicts trends in the improvement of life expectancy for selected male and female populations in developing countries. Most of the countries selected show almost linear improvement over the past 40 years, with several gaining more than 20 years of life expectancy. Cuba shows levels of life expectancy that are higher – at least for men – than in the United States and some European countries. Probably the most impressive increase (about 25 years) occurred in China. India also had a very substantial increase from less than 40 years in 1950 to around 60 years in 1990. A remarkable feature of Indian mortality is that until about 1985

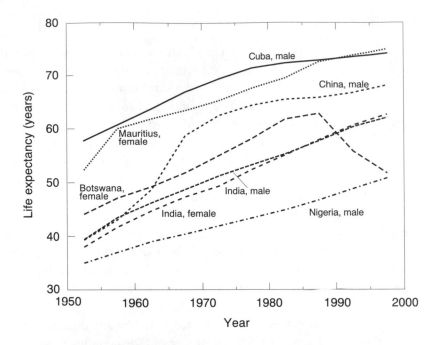

Figure 2.4. Trends in life expectancy at birth of selected male and female population in developing countries since 1950. Source: United Nations, 1997.

male life expectancy was higher than female life expectancy, an unusual pattern that is likely due to extreme discrimination against women and girls in Indian society. Generally, *Figure 2.4* gives the impression of striking increases in life expectancy all around the developing world (see also *Table 2.1*). An exception is Botswana, which is already beginning to experience the impacts of extremely high HIV prevalence.

As with fertility, uncertainties about future mortality trends are highest for sub-Saharan Africa, partly due to the lack of reliable data. For some countries in the region, available data show that infant and child mortality declines may have stalled, although the observed trend could be due to poor data quality (Bucht, 1996). There is no doubt that modern health infrastructure developed significantly after 1960, as shown by an increasing number of health personnel per capita. Efforts were also made to introduce new vaccines and to build up primary health care strategies throughout the continent. Garenne (1996) concludes a survey of African mortality by stating that past trends were set in motion by transfers of technology from the West, which affected almost all countries in a short period of time. In the future, however, public health, nutrition, economic development, and education will be the key determinants of mortality decline. Concerning the future of African mortality, there are reasons to assume that the declining trend of the past 30 years will not continue and that differences between countries will persist or even increase. This discouraging picture is due to the spread of infectious diseases, especially HIV (see *Box 2.3*), as well as problems of basic subsistence.

In terms of the scenario assumptions, for sub-Saharan Africa the IIASA projections assume a three-year per decade decline in life expectancy as the low value, and a three-year increase as the high value, resulting in constant mortality for the central scenario. AIDS is also taken into consideration in the mortality assumptions for southern Asia and Southeast Asia, where the low case assumes zero improvements in mortality. The high values assume that AIDS can be largely brought under control and mortality will improve quickly following the path of today's industrialized countries.

More Developed Countries

Currently, life expectancy at the national level is highest in Japan at 82 years for women and 77 years for men. Only three decades ago, in 1960–1965, life expectancy was 72 for women and 67 for men, implying an average increase of more than three years per decade for Japanese women. Eastern Europe, on the other hand, had life expectancies similar to those of Japan during the early 1960s but has had little improvement (or even declines) since then, bringing life expectancy to about 74 for women and 63 for men at present (see *Figure 2.5*). Western Europe and North America take an intermediate position, with life expectancies that have increased steadily by about two to three years per decade. The recent trend points toward further improvements, and the analysis of age- and cause-specific mortality trends does not give indications of an imminent leveling off of improvements (Valkonen, 1991). For example, studies on occupational mortality differentials in the Nordic countries – which are more homogeneous than many other countries – still show significant inequalities, mostly by social class (Andersen, 1991), which may be taken as an indication of the possibility of further improvements if the higher mortality groups change their lifestyles.

Such considerations, and the rapid general improvements in mortality observed over the past several decades, have led institutions that produce population projections to repeatedly update their assumptions on possible future gains in life expectancy. In a series of steps the UN raised its assumed limit for life expectancy for men from 72.6 years (assumed in 1973) to 82.5 years (assumed in 1988). Similar adjustments were also made for women (Bucht, 1996).

The existence of a fixed biological limit to the human life span is, however, controversial. The more traditional view is that aging is a process intrinsic to all cells of the human body. Under this view, recent and possible future mortality improvements are interpreted as a "rectangularization" of the survival curve through the elimination of premature deaths and resulting concentration of mortality shortly

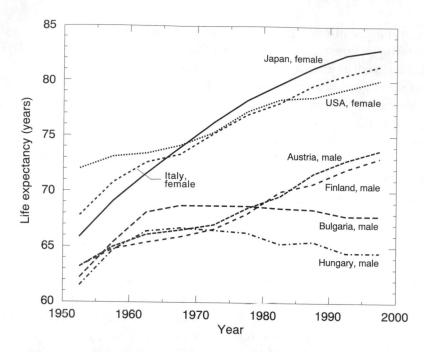

Figure 2.5. Trends in life expectancy at birth for selected male and female populations in industrialized countries since 1950. Source: United Nations, 1997.

before the maximum age. Based on this idea, a limit life table can be calculated with a maximum life span of 115 years and an average age of natural death of around 90 (Duchêne and Wunsch, 1991). This view implies lower rates of improvement in the future. Such lower rates may also be induced by worsening living conditions, health-impairing behaviors, or worsening environmental pollution (Day, 1991). Olshansky *et al.* (1990) also argue that an increase in life expectancy beyond 85 is unlikely because it would require a decline of more than 55% in mortality rates from all causes at all ages, which is improbable. Following these lines of reasoning, the IIASA scenarios assume an improvement in life expectancy of only one year per decade (for both men and women) as the low value.

  The alternative view sees aging as a multidimensional process of interaction in which partial loss of function in one organ can be compensated by other organs, and only total loss of a necessary organ system results in death (Manton, 1991). If conditions are favorable – through improved living conditions or possible direct intervention into the process of cell replication and aging – this could result in much higher average life expectancies in the future. There is not yet enough data to resolve this controversy, but Manton *et al.* (1991) present an impressive list of evidence from small special populations followed for short periods of time that show average life expectancies of well above 90 years. Further supporting evidence for this can be found by studying changes in very old age mortality over time (Vaupel *et al.*, 1998). Using Swedish data since 1900, Vaupel and Lundström (1996) show that mortality rates at ages 85, 90, and 95 have substantially declined at an accelerating rate, which over the past decades averaged 1–2% for women and 0.5% for men.

These considerations provided the motivation for the assumption of much more significant improvements in life expectancy in the IIASA high scenario – three years per decade – than those assumed by most other national and international population projections. The average of the high and low scenarios results in a central scenario improvement of two years per decade, which seems plausible as supported by Mesle (1993).[14]

2.4.4 Assumptions on future international migration

Recent immigration trends in Western Europe demonstrate the volatility of migratory streams. During the early 1970s, West Germany had an annual net migration gain of more than 300,000; this had declined to only around 6,000 five years later, and even to 3,000 during the early 1980s. During the 1985–1990 period, however, the annual net gain increased sharply again to 378,000, one hundred times that of the previous period. Few other countries show fluctuations as extreme as Germany's, but even the traditional countries of immigration – the United States, Canada, and Australia – show remarkable ups and downs. Annual net migration to Australia declined from 112,000 in the early 1970s to 54,000 in the late 1970s. During the 1980s it increased again to over 100,000. For the United States and Canada, data show an increase from around 280,000 per year during the early 1960s to around 800,000 in the mid-1990s.

Aside from the migration streams into Western Europe, North America, and Australia/New Zealand, labor migration within Asia has been remarkable over the past decades. For instance, during the early 1980s more than 330,000 workers left the Philippines annually, with around 280,000 of them going to the Middle East. The Republic of Korea lost an average of 170,000 workers per year during the early 1980s; India lost 240,000 and Pakistan 130,000 annually during the same period. The largest proportion of these workers went to oil-rich countries in the Middle East. During the late 1980s these migratory streams within Asia showed significant declines.

Due to the volatility of these trends and the great role that short-term political and economic changes play in both the receiving and sending countries, it is more difficult to project future migratory streams than fertility and mortality rates. Furthermore, net migration always results from the combination of two partly independent migration streams: that entering and that leaving a certain region for whatever reasons.

There is potential for further increases in interregional migration due to better communications between regions, cheap mass transport, and the persisting gap between the North, which is not only richer but also rapidly aging and in need of young labor, and the South, which has many young people with rising skill levels but low income opportunities in their home countries. In addition to these economic factors, expected environmental changes in many parts of the world may cause further significant pressure to migrate to other regions where conditions are perceived to be better. The issue of "environmental migration" is discussed in Chapter 5.

Still, the actual extent of future South–North migration streams depends not only on "pull" and "push" factors, but also on the immigration policies of the receiving countries. If the European Community, for instance, were to enforce a policy of closed outside borders, this might result in a situation of almost no net migration. Hence for all regions the low scenario value for net migration chosen was zero. This is not the same as assuming that borders will be closed to migrants; it only implies that, on average, the number of in-migrants approximately equals the number of out-migrants.

For the high migration scenarios, annual net migration gains of 2 million in North America, 1 million in Western Europe, and 350,000 in Japan/Australia/New Zealand have been assumed. For Eastern Europe special assumptions were made for the period 1990–2030, with out-migration dominating the first decade.

The distribution of migrants from the assumed sending regions to the receiving regions and migration patterns among the developing regions are based mostly on currently observed migratory streams as given in Zlotnik (1996). Annual net migration is assumed to be –475,000 in northern Africa, –510,000 in sub-Saharan Africa, –650,000 in Central America, –260,000 in South America, –15,000 in western Asia, –835,000 in southern Asia, –470,000 in China, and –1,135,000 in Southeast Asia. Model migration schedules by Rogers and Castro (1981) were used to determine the age patterns of migrants.

2.5   The IIASA 1996 Population Projections: Results

Using the high, central, and low scenarios for fertility, mortality, and migration in 13 world regions, a number of projections of future population were performed. We summarize results of two distinct approaches. More detailed results can be found in Lutz (1996). In the first approach, alternative combinations of assumptions for vital rates (i.e., mortality and fertility) give rise to a family of alternative scenarios for future population and age structure. A comparison of these scenarios allows conclusions to be drawn regarding the sensitivity of results to the underlying assumptions. In the second approach, the high–low range for each vital rate is interpreted as a 90% uncertainty interval and a set of probabilistic projections is performed, resulting in a distribution of outcomes for population size and age that provides new insight into the likelihood of demographic trends over the present century.

2.5.1 Alternative scenarios

Table 2.3 gives the population sizes for 2020, 2050, and 2100 for the nine scenarios resulting from the combination of the three fertility assumptions (low, central, and high) with the three mortality assumptions. Migration is held at its central level in all of these scenarios. Results are presented at the level of major world regions.

The combination of central fertility with central mortality assumptions is considered the most likely scenario. If these central assumptions hold for all world regions then the total world population will increase from 5.7 billion in 1995 to 7.9 billion in 2020 and to 9.9 billion in 2050. An extension to the second half of the century then yields a leveling off of world population growth, reaching an estimated total population size of about 10.4 billion by 2100. Under this scenario, today's MDCs will see a moderate increase of about 90 million by 2020, followed by a moderate decline. The population of the LDCs, on the other hand, will more than double.

Table 2.3 also gives information about all other combinations of fertility and mortality in the different world regions. The lowest population path results from combining high mortality with low fertility in every world region. In this case, world population levels off during the 2030s, reaches 7.1 billion in 2050, and by 2100 declines to about 4 billion. The highest population path combines low mortality with high fertility, producing a world population of 13.3 billion by 2050 and 22.7 billion by 2100.

However, these extremes are considered unlikely, not only because they are based on maximum or minimum values of a range judged to cover 90% of all possible cases, but also because they suffer from an internal contradiction. For example, if fertility remains high in LDCs, it is likely to be a result of slow economic and social progress, in which case it is more likely that mortality will be high, not low. Similarly, if rapid change results in low fertility, it is more likely that mortality will be low, not high. Policy interventions, as well, are more likely to lead to combinations of low mortality and low fertility, or high mortality and high fertility, than to the low–high and high–low combinations which give rise to the most rapid and slowest rates of population growth. For example, policies designed to accelerate fertility decline also tend to lead to mortality improvements. Scenarios corresponding to these more plausible combinations of assumptions might be called the "rapid demographic transition" scenario (low mortality and low fertility) and "slow demographic transition" scenario (high mortality and high fertility).

Table 2.3. Projections of total population (in millions) according to alternative fertility and mortality assumptions (and central migration assumptions) for 2020, 2050, and 2100.

Mortality	1995	Fertility Low 2020	Low 2050	Low 2100	Central 2020	Central 2050	Central 2100	High 2020	High 2050	High 2100
High										
Africa	720	1,202	1,312	707	1,278	1,710	1,484	1,355	2,173	2,739
Asia-East	1,956	2,239	1,979	994	2,412	2,601	2,308	2,589	3,369	4,650
Asia-West	1,445	2,073	2,210	1,241	2,181	2,741	2,432	2,290	3,352	4,321
Europe	808	769	593	291	811	716	546	854	860	961
Latin America	477	644	687	454	686	867	927	729	1,080	1,700
North America	297	335	321	251	354	386	423	374	463	697
Less developed	4,451	6,011	6,071	3,330	6,404	7,780	7,040	6,802	9,810	13,228
More developed	1,251	1,250	1,032	607	1,319	1,241	1,081	1,388	1,486	1,841
World	5,702	7,261	7,103	3,937	7,723	9,021	8,120	8,191	11,300	15,070
Central										
Africa	720	1,252	1,562	1,140	1,332	2,040	2,366	1,414	2,599	4,349
Asia-East	1,956	2,268	2,116	1,205	2,444	2,760	2,704	2,621	3,554	5,348
Asia-West	1,445	2,117	2,421	1,634	2,228	2,995	3,136	2,338	3,659	5,508
Europe	808	782	640	342	825	766	624	868	912	1,075
Latin America	477	650	722	530	693	906	1,056	737	1,125	1,910
North America	297	339	340	283	359	406	467	378	483	757
Less developed	4,451	6,138	6,694	4,433	6,541	8,554	9,137	6,946	10,765	16,915
More developed	1,251	1,270	1,107	700	1,340	1,319	1,216	1,410	1,568	2,033
World	5,702	7,408	7,802	5,134	7,879	9,874	10,350	8,356	12,330	18,950
Low										
Africa	720	1,298	1,799	1,662	1,381	2,344	3,348	1,467	2,983	6,058
Asia-East	1,956	2,296	2,257	1,440	2,473	2,918	3,090	2,652	3,733	5,960
Asia-West	1,445	2,158	2,624	2,073	2,270	3,235	3,847	2,383	3,939	6,627
Europe	808	796	692	402	839	819	705	882	968	1,185
Latin America	477	656	756	612	700	945	1,180	743	1,169	2,096
North America	297	344	360	318	363	427	511	383	505	812
Less developed	4,451	6,257	7,301	5,701	6,665	9,285	11,325	7,080	11,643	20,524
More developed	1,251	1,291	1,188	806	1,360	1,402	1,355	1,431	1,654	2,214
World	5,702	7,547	8,488	6,507	8,026	10,690	12,680	8,510	13,300	22,740

Discrepancies in totals are due to rounding.

Source: Lutz, 1996.

These less extreme (from the standpoint of population growth) but more meaningful scenarios are given in the upper right and lower left corners of *Table 2.3*. In the first case, world population rises to 15.1 billion in 2100; in the second case, it rises to 8.5 billion in 2050 before declining slowly to 6.5 billion in 2100. When in the following chapters we discuss the sensitivity of GHG emissions to population trends, we compare the central scenario with the rapid demographic transition scenario, which we assume can be brought about by appropriate policies affecting population.

These global totals hold only if fertility and mortality trends are perfectly correlated across all world regions (for example, all regions simultaneously follow the high fertility assumptions). In the real world, however, it is much more likely that one region will follow one scenario while another follows an alternative scenario. Why, for example, should the fertility trend in Latin America be correlated to that in Eastern Europe? *Table 2.3* allows the user to define any mix of scenarios in different world regions and aggregate them into global population sizes. All these regionally mixed scenarios will lie between the extremes because diverging trends in different regions will partly offset each other.

Table 2.4 gives results for population aging, here indicated by the proportion of the population above age 60. From a comparison of *Tables 2.3* and *2.4*, an important generalization becomes apparent: the lower the fertility (and therefore population growth), the older the projected population age structure; and the higher the fertility, the younger the projected population age structure. But the table also shows that significant population aging is inevitable. Even in the slow demographic transition scenario combining high fertility with high mortality (which also tends to reduce aging), the proportion above age 60 increases from the current 9.5% to 17.8%. In the rapid demographic transition scenario combining low fertility with low mortality, the proportion above age 60 increases to 42.5% of the total population. Considered in terms of increases in absolute size of the population above age 60, rather than as increases in the proportion, the rapid and slow demographic transition scenarios, as well as the central scenario, imply a fivefold increase in the 60+ age group in the long run.

Deceleration of population growth is a broadly supported goal of international population policy. The estimates in *Tables 2.3* and *2.4* reveal, however, two qualifications that should be kept in mind. First, the more rapidly fertility declines, the more population aging is accelerated and intensified. Second, as illustrated in *Table 2.3*, the slowest population growth is associated with the highest mortality, which is hardly a desirable option. On both counts, population stabilization is not a goal that should be pursued blindly.

Table 2.4. Projections of percentages of the population above age 60 according to alternative fertility and mortality assumptions (and central migration assumptions) for 2020, 2050, and 2100.

| | | Fertility | | | | | | | | | |
| | | Low | | | Central | | | High | | |
Mortality	1995	2020	2050	2100	2020	2050	2100	2020	2050	2100
High										
Africa	5.0	6.2	12.5	28.5	5.8	9.6	20.0	5.5	7.5	14.8
Asia-East	9.4	16.1	29.7	34.9	15.0	22.6	24.6	14.0	17.4	18.0
Asia-West	6.6	9.9	18.5	34.1	9.4	14.9	24.4	9.0	12.2	18.2
Europe	17.8	25.1	38.1	37.5	23.8	31.6	28.5	22.6	26.3	22.0
Latin America	7.6	12.4	24.4	33.1	11.6	19.4	23.8	10.9	15.5	17.8
North America	16.4	24.5	32.9	33.2	23.1	27.4	26.3	21.9	22.8	20.6
Less developed	7.2	11.2	21.1	32.9	10.5	16.4	23.3	9.9	13.0	17.3
More developed	17.7	25.7	36.9	36.2	24.4	30.7	28.1	23.1	25.7	21.9
World	9.5	13.7	23.4	33.4	12.9	18.4	24.0	12.2	14.7	17.8
Central										
Africa	5.0	6.3	13.1	30.7	6.0	10.0	21.5	5.6	7.9	15.8
Asia-East	9.4	16.6	31.9	40.1	15.4	24.5	28.4	14.3	19.0	20.9
Asia-West	6.6	10.1	19.7	37.7	9.6	16.0	27.2	9.2	13.1	20.3
Europe	17.8	25.9	41.4	44.1	24.6	34.6	34.0	23.4	29.1	26.3
Latin America	7.6	12.7	26.0	37.5	11.9	20.7	27.1	11.2	16.7	20.3
North America	16.4	25.2	35.9	39.1	23.8	30.0	31.2	22.6	25.2	24.6
Less developed	7.2	11.5	22.2	36.4	10.8	17.4	25.9	10.1	13.8	19.2
More developed	17.7	26.5	40.2	42.5	25.1	33.7	33.4	23.9	28.4	26.1
World	9.5	14.0	24.8	37.2	13.2	19.6	26.8	12.5	15.7	19.9
Low										
Africa	5.0	6.5	14.3	35.8	6.1	11.0	25.2	5.8	8.6	18.5
Asia-East	9.4	17.0	34.5	46.6	15.8	26.7	33.3	14.7	20.9	24.3
Asia-West	6.6	10.4	21.3	43.0	9.9	17.3	31.2	9.4	14.2	23.4
Europe	17.8	26.8	44.9	51.1	25.4	38.0	40.0	24.2	32.1	31.1
Latin America	7.6	13.0	27.8	42.8	12.2	22.3	31.1	11.5	18.0	23.2
North America	16.4	25.9	38.9	45.0	24.5	32.9	36.1	23.2	27.8	28.5
Less developed	7.2	11.7	23.8	41.6	11.0	18.7	29.8	10.4	15.0	22.1
More developed	17.7	27.3	43.6	49.1	25.9	36.9	39.0	24.7	31.3	30.7
World	9.5	14.4	26.6	42.5	13.5	21.1	30.8	12.8	17.0	22.9

Source: Lutz, 1996.

Sensitivity

Fertility assumptions, as we said at the beginning of this section, have a far greater impact on projected long-term population size than do mortality assumptions. Holding fertility and international migration assumptions at their central values and allowing only mortality assumptions to vary, the population forecast range for 2100 is between 8.1 billion in the high mortality case and 12.7 billion in the low mortality case (see *Table 2.3*). Holding international migration and, this time, mortality assumptions at central values and allowing only fertility to vary, the range of variation is between 5.1 billion in the low fertility scenario and 19.0 billion in the high fertility scenario. Considered region by region, population size is more sensitive to the future course of fertility in LDCs than in MDCs, where the range of possible variation in behavior is lower and the base is smaller.

Fertility is also the most important factor governing population aging. Under central mortality and migration assumptions, the high fertility scenario implies 19.9% of the population over age 60 in the year 2100 and the low fertility scenario implies 37.2% – a spread of over 17 percentage points (see *Table 2.4*). Under central fertility and migration assumptions, however, the high mortality scenario implies 24.0% of the population over age 60 and the low mortality scenario implies 30.8%, a spread of only about 7 percentage points. Migration assumptions have even less of an impact on aging than mortality. Taking Western Europe as an example, if mortality and fertility assumptions are held at central levels and migration assumptions are varied, the forecast range for the proportion of the population over age 60 in 2100 is only between 34.1% in the high migration case and 36.5% in the low migration case. In terms of absolute change in the proportion of the population over age 60, MDCs and LDCs are about equally sensitive to the future course of fertility. For example, in 2050 the difference between the low and central fertility scenarios (given central mortality and migration assumptions) is 6.5 percentage points (40.2% as opposed to 33.7%) for MDCs and 4.8 percentage points (22.2% as opposed to 17.4%) for LDCs.

Projecting Numbers of Households

As discussed in Chapter 4, studies of the environmental impact of demographic change must choose a particular demographic unit of account on which to base analyses. For many applications, including the role of population variables in GHG emissions, it can be argued that the number of households is at least as important as the number of people, since significant economies of scale in energy use exist at the household level. It is therefore useful to be able to project trends in numbers of households that are consistent with underlying population projections.

Table 2.5. Projections of numbers of households for MDCs and LDCs, 1990–2100, used in the GHG emissions analysis in Chapter 4.

Scenario	Population (millions)		Households (millions)		Average household size	
	MDCs	LDCs	MDCs	LDCs	MDCs	LDCs
Central						
1990	1,266	3,986	469	830	2.70	4.80
2020	1,428	6,453	544	1,623	2.63	3.98
2050	1,457	8,416	567	2,465	2.57	3.41
2100	1,411	8,942	564	3,312	2.50	2.70
Rapid Demographic Transition						
1990	1,266	3,986	469	830	2.70	4.80
2020	1,370	6,178	554	1,647	2.48	3.75
2050	1,289	7,200	544	2,312	2.37	3.11
2100	899	5,608	391	2,337	2.30	2.40
Slow Demographic Transition						
1990	1,266	3,986	469	830	2.70	4.80
2020	1,485	6,705	535	1,596	2.78	4.20
2050	1,670	9,626	596	2,592	2.80	3.71
2100	2,202	12,867	786	4,289	2.80	3.00

Note: Definitions of MDCs and LDCs have been modified from those of Lutz (1996) by moving Central Asia from the LDCs to the MDCs to agree with the regional definitions used by the IPCC emission scenarios. Population totals for 1990 are from Pepper *et al.* (1992), also for consistency with the IPCC scenarios.

Trends in number of households may be broken down into two components: the trend in age-specific household headship rates, defined as the number of heads of household in a specific age group divided by the total size of the age group, and changes in the age structure of the population. The first trend reflects behavioral change; MDCs, for example, have undergone a pronounced shift toward higher rates of household headship (and therefore lower average household size) with the decline of the extended family. The second factor is an accounting effect: population aging increases the weight of age groups with high headship rates (since the elderly are more likely to live alone) and reduces the weight of age groups with low headship rates (since children and young persons are unlikely to be household heads). This leads to an increase in the number of households independent of changes in headship rates. MacKellar *et al.* (1995) demonstrate that the accounting effect due to aging is likely to be more important than possible changes in headship rates over this century. They produce projections of numbers of households based on the IIASA central, rapid demographic transition, and slow demographic transition scenarios using age-specific headship rates held constant in all regions at current levels.

Results for MDCs and LDCs are given in *Table 2.5*. They show that the number of households in both regions will grow much faster than the number of people. For example, in the central scenario, LDC population slightly more than doubles between 1990 and 2100. Due to aging, however, average household size is reduced nearly by half over the same period, so that the number of LDC households grows by about a factor of four. This increase would be accentuated if, as seems reasonable, household headship rates were assumed to converge with those in MDCs. In MDCs, population grows about 11% by 2100, while the number of households grows about 20% due to a slight decline in the average household size (from 2.7 to 2.5).

2.5.2 Probabilistic projections

There is no one simple way to apply probabilistic methods to population projections. Assumptions about the future variance of the distributions of the three components have traditionally been based on time series analysis or *ex post* analysis of projection errors. Both approaches have problems, but the most important flaw for a global projection is the lack of appropriate time series data for large parts of the world population. For this reason, the IIASA projections use an approach that is based on expert judgment. The procedure fits a normal distribution to the three values (high, central, and low) that resulted from the expert discussions, with 90% of the cases lying between the high and low values.[15] Results were derived through a set of 4,000 simulations that randomly combined fertility, mortality, and migration paths from the three normal distributions for the 13 world regions. These simulations also considered the possibility that fertility and mortality trends may be correlated within regions (e.g., high fertility in sub-Saharan Africa is more likely to go hand in hand with high mortality than low mortality) and that regional trends may be either independent of each other (e.g., fertility in sub-Saharan Africa uncorrelated with fertility in Latin America) or correlated.

Figure 2.6 shows the distribution of future population sizes derived from the full set of 4,000 projections at five-year intervals to the year 2100. The high and low boundaries give the range into which 95% of all cases fall. The upper line indicates that there is an unlikely possibility of almost linear population growth between 1995 and 2100. The lower line shows that there is also an equally unlikely possibility that the world's population will peak in the middle of the 21st century and fall thereafter to below 6 billion by 2100. The much more probable range of future paths (between the 0.2 and 0.8 fractiles, covering 60% of all cases) is remarkably small. By 2050 this uncertainty range is less than 1.5 billion people, and by 2100 it doubles to about 3 billion people.

The figure also shows that in more than 60% of all cases total world population levels off during the second half of this century, or even starts to decline. Given that

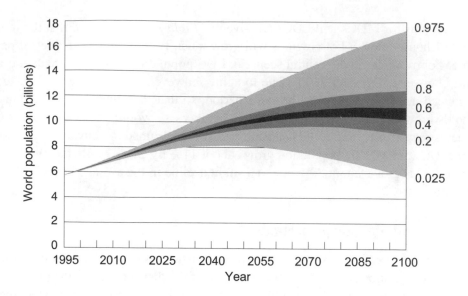

Figure 2.6. Merged distribution of the size of the world's population (in billions), 1995–2100. Source: Lutz *et al.*, 1996a.

the world population in mid-1996 was estimated at 5.8 billion, Lutz *et al.* (1996a, 1997) find that the probability of a doubling – that is, reaching or surpassing the 11.6 billion mark at any point during this century – is only 33%. In other words, there is a two-thirds probability that world population will not double by the end of the 21st century.

Regional results for the case of uncorrelated fertility and mortality trends are summarized in *Table 2.6*. Sub-Saharan Africa displays the largest range of uncertainty in future population size, with a 95% confidence interval in 2100 ranging between 578 million and 4.345 billion around a mean of 1.9 billion. This large range results from the unusually large uncertainty surrounding trends in fertility and mortality in the region, in addition to the assumption that the two trends will be uncorrelated. Lutz *et al.* (1996a) have shown that if fertility and mortality are instead assumed to be positively correlated within the region, the range of uncertainty becomes significantly smaller.

Figure 2.7 shows the distribution of the percentage of the global population over age 60. All lines are rising, indicating near certainty that the percentage of older people in the population will rise over time. In 2050, the median is 20% (with a 95% uncertainty interval between 15% and 26%), compared with 9.5% in 1995. By 2100, the median is 27%, with a 95% uncertainty range of 19% to 41%. In other words, there is a 95% chance that the proportion of elderly will at least double over the course of this century. In the most likely case it will almost triple,

Table 2.6. Population (in millions) by region for probabilistic projections assuming uncorrelated fertility and mortality. Mean, median, and 95% confidence intervals for 2020, 2050, and 2100.

Region	1995	2020 Mean[a]	2020 Med.[b]	2020 2.5%[c]	2020 97.5%[c]	2050 Mean[a]	2050 Med.[b]	2050 2.5%[c]	2050 97.5%[c]	2100 Mean[a]	2100 Med.[b]	2100 2.5%[c]	2100 97.5%[c]
North Africa	162	277	277	254	300	440	439	309	583	630	598	228	1,202
Sub-Saharan Africa	558	1,059	1,058	965	1,159	1,625	1,605	1,085	2,316	1,909	1,738	578	4,345
China & CPA[d]	1,362	1,670	1,670	1,526	1,826	1,888	1,865	1,351	2,574	2,051	1,873	709	4,428
Pacific Asia	447	629	629	576	678	802	796	579	1,047	876	829	322	1,696
Pacific OECD	147	155	155	145	167	146	146	117	182	125	120	59	221
Central Asia	54	87	87	76	100	139	137	88	206	212	194	65	477
Middle East	151	300	300	279	324	520	515	380	692	786	738	320	1,516
South Asia	1,240	1,845	1,845	1,737	1,949	2,380	2,368	1,833	2,970	2,365	2,246	1,014	4,327
Eastern Europe	122	124	124	116	133	111	110	86	141	83	78	31	168
European FSU[e]	238	224	224	209	240	189	188	144	241	147	138	53	290
Western Europe	447	479	479	446	512	472	471	370	584	430	416	196	769
Latin America	477	697	696	646	746	930	925	707	1,177	1,163	1,106	489	2,142
North America	297	356	356	320	400	405	403	303	534	482	467	229	865

[a]Data on the mean population size.
[b]Data on the median population size.
[c]Columns labeled 2.5% and 97.5% provide data on the lower and upper bounds, respectively, of the 95% confidence interval; 2.5% of all observations lie below the lower bound and 97.5% of all observations lie below the upper bound. All figures are based on 1,000 simulations and were produced using DIALOG, the multistate population projection model. Fertility and mortality are assumed to be uncorrelated within regions.
[d]China and centrally planned Asia.
[e]European part of the former Soviet Union.
Source: Lutz, 1996.

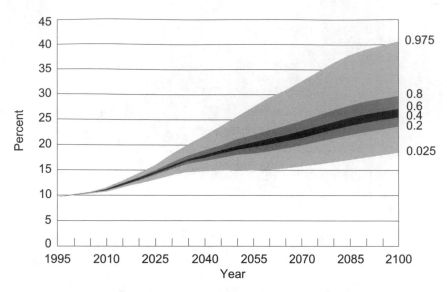

Figure 2.7.   Merged distribution of global population over 60 (in percent), 1995–2100. Source: Lutz *et al.*, 1996a.

and it may even more than quadruple. The uncertainty in the percentage above age 60 grows significantly during the second half of the 21st century due mainly to the uncertainty in future old-age mortality when combined with uncertainty in fertility.

To summarize, three near certainties emerge from the range of scenarios and from the probabilistic projections.

- World population will increase substantially from its current level. Even in the lowest growth scenario, population increases by close to 2 billion before commencing its decline. However, a doubling of world population has become unlikely in the face of recent fertility declines.
- The distribution of world population will continue to shift toward LDCs. Even assuming rapid fertility decline and little improvement in mortality, LDCs still account for a rising share of world population.
- The world population will continue to age. Probabilistic projections show that a doubling of the proportion above age 60 is a near certainty.

2.6   Impact of Global Climate Change on the Population Outlook

Climate change is likely to have contributed to past regional declines in population size, such as the collapse of the Classic Maya culture in Yucatán (A.D. 800–1000;

Lutz *et al.*, 1996b), the decline of the Easter Island civilization (Brander and Taylor, 1998), and the abandonment of Greenland by early Norse settlers (ca. A.D. 1000; O'Hara *et al.*, 1993). On the other hand, these massive disruptions occurred in a context of extremely limited technical capacity to respond to change (the first two cases) or simply represent migration on a large scale (the last case). In other cases where the human population has been found to be sensitive to climate, effects have been modest (Lee, 1987; Galloway, 1986; Richards, 1983). Although significant climate change has occurred at times in particular regions, it has never seriously threatened the existence of the human species.

This will almost certainly continue to be true. A simulation based on the IIASA scenarios (Lutz, 1994) shows that even extreme mortality crises would not have a lasting, decisive impact on population size. A projection was performed which assumed that in the middle of the 21st century an eruption of starvation, disease, and war kills 20% of the population of sub-Saharan Africa over the course of a few years. A demographic recovery quickly asserts itself; within 10–15 years, population size returns to its previous level.[16] It turns out that only a mortality crisis that is selective of reproductive-aged females would have the potential to reduce human numbers drastically. In fact, females aged 15–45 are probably the most robust members, in terms of health, of the human population. Nothing resembling such a catastrophe has appeared in the literature on projected impacts of global climate change (see Chapter 5).

Exogenous shocks will also be mediated through the social side of human demographic response. Lee (1990) studied the demographic response to a run of bad harvests in agricultural communities. In a typical case, vulnerable members of the population, particularly the aged, exhibit increased frailty and die as a result of normal morbidity conditions. Men may leave the community to seek work in adjoining agricultural regions or in cities; the ensuing spousal separation leads to a reduction in fertility. As financial assets are deployed to purchase food, marriages are delayed, reducing the number of first births. If there is a shortage of weaning supplements, infants will be breast-fed more intensively for a longer duration, with consequent reduction in fecundibility. Davis (1963) long ago coined the term "multiphasic response" to refer to this dynamically complex demographic reaction to stress. Expectations and the incremental nature of many crises also play a role; for example, Dyson (1991) found an anticipatory demographic response in South Asia, where famines were preceded by extended periods of worsening adversity during which fertility declined.

Moving from preindustrial to modern societies, policy responses mediate between exogenous shocks and demographic impacts. While research can give some estimate of potential impacts of climate change, actual impacts will be a function of potential impact, uncoordinated human responses, and human responses coordinated through policies. As discussed in more detail in Chapter 5, climate change impacts on agriculture may be serious in particular regions, but the potential for adaptation and the probability that global agricultural output can be maintained, make a large-scale demographic catastrophe unlikely.

The group of experts that defined the low, central, and high mortality assumptions of the projections discussed above were explicitly instructed to include in the range of possible future mortality trends all environmental threats as we can foresee them today. Hence the high mortality assumptions in the above scenarios explicitly include possible negative effects of future environmental problems. Extremely unlikely cataclysmic events under this assessment can be considered to occupy the 5% tail on the high end of the assumed probability distribution for mortality.

Similarly, possible intercontinental population movements induced by global environmental change were kept in mind when defining the low, central, and high migration assumptions. The high migration case implicitly describes a world of very strong migration to the North due to serious problems in the South. In Chapter 5, more attention is given to environmental factors in migration.

2.7   Conclusion

This chapter has identified three near certainties in the world population outlook: further population growth, further tilt toward LDCs in the world distribution of population, and significant population aging. These robust findings follow from a wide range of extreme scenarios as well as a probabilistic population projection exercise, both embodying the sectoral expertise of an international panel of researchers.

The findings support the conventional wisdom that there is one key uncertainty from the standpoint of the future size and age structure of the total human population: the speed of fertility decline in LDCs and, in the longer term, the ultimate fertility level reached. Our understanding of the factors influencing fertility is far from complete. However, research supports the view that a combination of effective national family planning programs, socioeconomic development, and empowerment of women results in fertility decline. Because it positively contributes to all three, investing in female basic education is likely the policy intervention that most strongly accelerates fertility decline.

While less important than fertility, future mortality trends will also have a significant impact on future global population size and age structure. Although according to current understanding climate change may increase mortality in some regions, it is unlikely to affect total human numbers drastically.

It seems safe to predict, based on the projections summarized in this chapter, that the international population community, as well as policymakers, will gradually shift their concern away from population size and rate of growth to population age structure, specifically, aging. While the MDCs are entering unknown territory in terms of the proportion of the population over age 60, the LDCs are approaching the age structure of today's MDCs. Thus, under the central scenario, the proportion of the population over age 60 in MDCs is expected to rise from 17.7% in 1995 to 33.7% in 2050 while in LDCs it will rise from 7.2% to 17.4%. Based on the sensitivity analyses presented here, the extent of population aging is equally sensitive to choice of fertility scenario in the two regions.

Notes

[1] The total fertility rate (TFR) is defined as the number of children a woman would give birth to if through her lifetime she experienced the set of age-specific fertility rates currently observed. Since age-specific rates generally change over time, TFR does not in general give the actual number of births a woman alive today can be expected to have. Rather, it is a synthetic index meant to measure age-specific birth rates in a given year.

[2] Life expectancy at birth is the average lifespan implied by the observed age-specific mortality rates in a given year. Like the TFR, it is useful as a snapshot of prevailing conditions; however, because it is not specific to a particular cohort, it does not indicate the lifespan expected for a newborn.

[3] Replacement level fertility is the TFR that, if maintained over the long term, would result in an equilibrium in which each generation exactly replaces itself. Replacement level TFR depends on the sex ratio at birth (which varies slightly from population to population) and mortality between infancy and the end of the reproductive life span. In countries that have undergone the initial mortality decline stage of the demographic transition, the replacement level TFR is roughly 2.1.

[4] Differences in pretransition fertility rates also contributed to differences in population growth rates between MDCs and LDCs. Due to typical European marriage patterns (high age at marriage and high proportion remaining unmarried), MDC fertility began the transition at a lower level than did LDC fertility.

[5] "Marriage" is a culturally defined word, but the proximate determinants framework has proved sufficiently flexible to become the standard model in accounting for changes in fertility rates. Variables such as incidence of spousal separation, frequency of sexual intercourse, and natural fecundibility have not been found to vary sufficiently to be of much concern.

[6] In the initial phase of fertility decline an increase in the age at first marriage also played an important role in some countries.

[7] The neoclassical model of population, economic development, and environment is discussed in more detail in Chapter 3.

[8] A multiple regression analysis employing data for 37 WFS countries shows that, after controlling for other socioeconomic variables, one additional year of schooling is associated with a reduction in desired family size of 0.1 children (United Nations,

1987, p. 246). This simple correlation masks what is likely a nonlinear relationship. For example, fertility may be higher for women with low education than for those with no education because women with some education are likely to be healthier. At higher levels of educational attainment, the inverse correlation between schooling and fertility asserts itself. Cross-sectional results linking female education and both desired and actual family size are consistent with a large body of household-level studies done within countries.

[9] While the correlation in LDCs between child survival and the TFR is stronger than that between educational attainment and the TFR (United Nations Secretariat, 1989, p. 100), education is causally linked to child survival while the reverse is not true.

[10] Another possible link between women's status and low fertility might be through improved health of children.

[11] Migration rates are relevant as well, because they change the distribution of population between low and high fertility and mortality regions, but this is a second-order effect.

[12] Specific alternative fertility and mortality assumptions are given in Appendix I.

[13] The high assumptions are 2.3 in North America and Eastern Europe, and 2.1 in Western Europe and Japan/Australia. The assumptions are slightly higher for North America because of its different ethnic composition.

[14] For Eastern Europe a broader margin was chosen because of greater uncertainties. A lower value of 0.5 years was taken, assuming a continuation of the past slow improvements; 4.0 was taken as the upper value because there is a greater potential to catch up with the West if lifestyles became similar to the Western pattern.

[15] Sensitivity analyses assuming that the high and low values covered 85% and 95% instead of 90% of all cases showed that the results are relatively insensitive to this parameter. Thus the high–low range may be thought to cover 85–95% of all cases. The expert views are documented in Lutz (1996).

[16] The scenario assumes an agricultural carrying capacity of 2.5 billion people, derived by taking the mean of two scenarios from a joint FAO/UNFPA/IIASA (1982) land carrying capacity study. The first estimates a capacity of 4 billion people assuming medium agricultural inputs; the second estimates a capacity of less than 1 billion assuming low inputs.

Chapter 3

Population, Economic Development, and Environment

The two primary links between population and climate change that we examine in Part II – population's role in generating greenhouse gas (GHG) emissions and its role in affecting the ability of societies to adapt to the expected impacts of climate change – run through economic systems. Additionally, many of the justifications for population policy that we review in considering the policy implications of those links are based on economic effects. This chapter is intended to outline mainstream thinking in the field of population–economy–environment interactions in order to provide a framework for that analysis.

The population and economic development literature is extensive, and this chapter can at best only summarize a few basic themes. We first develop a basic neoclassical economic growth model, discussing in particular relationships between population growth and economic growth. We then add the environment link to form a neoclassical population–economy–environment model; we focus in particular on low-income settings, as it is these populations that are most vulnerable to the anticipated impacts of climate change. We examine the influence of population on the environment within the neoclassical model and contrast this with the view of ecological economics.

The population–environment model that predominates in the current economic literature and that has become influential in policy formulation is a basic neoclassical model extended to encompass poverty and low status of women and children. This model emphasizes how vulnerable populations react to environmental stress and how those reactions can set in motion a destructive downward spiral, or vicious circle.

The closing section looks at recent research on the macroeconomic impacts of population aging. Impacts of aging on productivity are not well understood.

Much research suggests that population aging will place downward pressure on saving; on the other hand, the accompanying slowdown in labor-force growth will simultaneously reduce the need for investment.

3.1   Population and Economic Development

The interpretation of population in economic rather than purely political or military terms is a recent development in social thought. Both Plato (*The Laws*) and Aristotle (*The Politics*) argued that the population of the city state should be limited, but this was to avoid problems of governance arising from large populations. Roman authors regarded population in terms of the availability of soldiers and the maintenance of the elite class. As early as the reign of Augustus, low rates of marriage and fertility among the Roman aristocracy were a source of concern, leading to the propagation of laws to encourage unions among the nobility.

With the depopulation of Europe during the Dark Ages and the Black Death of the 14th century as background, it is not surprising that medieval writings stress the positive side of population growth. This strand of thought persisted through the Renaissance and found outlet in the mercantilist school, which emphasized the positive role of population growth in stimulating commercial demand. This view was expressed most succinctly by the mercantilist economist Jean Bodin in 1576: "There is no wealth but men."

3.1.1   Malthus

The classical economics of Adam Smith and Thomas Malthus held a dimmer view of the link between population and the economy. In *The Wealth of Nations*, Smith wrote that the law of diminishing returns was likely to regulate population growth through wage rates. When population was large relative to fixed resources, wages would be low, discouraging marriage and reducing child survival rates. On the other hand, when labor was scarce, rising wages would stimulate marriage and favor the survival of children, and population would grow.

Malthus, in the first (1798) edition of his *Essay on the Principle of Population*, argued that the dynamics Smith described would drive wages to the subsistence level, where workers would be trapped in poverty. Expressed in terms of food, the Malthusian model held that while population requirements increase linearly with population size, production increases less than linearly because of the diminishing marginal product of labor on a fixed land base. Echoing Smith's wage feedbacks, the model predicted that if the population fell below its equilibrium level, production would exceed requirements and the population would grow; if population grew beyond its equilibrium level, requirements would exceed production and the population would diminish. Malthus postulated that wars, famine, disease, and other responses would serve as the "positive checks" on a large population. If the population was too small, then improved infant survival rates and "passion between the sexes" – earlier marriage and, to a lesser extent, higher marital fertility – would cause it to increase.

The model of Malthus, in its simplicity and clarity, is among the most compelling in the social sciences. However, it has proved to be lacking in several respects. Nineteenth century Europe, for example, escaped the Malthusian nightmare in three ways. First, the land base proved not to be fixed – Malthus never foresaw the opening up of the vast agricultural plains of North America and other regions. Second, despite the pervasive changes in agricultural technology that were going on around him, Malthus underestimated the possibilities for substituting capital for land and labor. Similarly, he failed to anticipate improvements in storage facilities and transport networks. Finally, Malthus had no way of predicting the Western fertility decline of the 19th and early 20th centuries.

3.1.2   A neoclassical economic growth model

Neoclassical economics provides two of the key components missing from the Malthusian model. First, unlike land in the Malthusian model, the pool of physical capital is not fixed; rather, it is expandable through saving and investment. If the capital stock expands faster than the labor force grows (known as "capital deepening"), the capital-to-labor ratio will increase, leading to growth in per capita output. Malthus' "positive checks" are thereby avoided. It can be shown that in order for capital deepening to occur, the investment rate, assumed in a closed economy to be equal to the proportion of output saved (s), must be greater than the labor-force growth rate (r) multiplied by the ratio of capital to output (k); that is, $s > rk$.[1]

Given a particular saving rate, a change in the population growth rate (taken here as roughly equivalent to the growth rate of the labor force) changes the equilibrium level of per capita output. Suppose population growth accelerates; given a fixed saving rate, capital will be spread more thinly over the labor force (known as "capital dilution") and per capita output will fall to an equilibrium level that is just sustained by the saving rate at the new rate of population growth. The model therefore predicts an inverse relationship between the population growth rate and the level of per capita output. However, given reasonable assumptions regarding the ease with which capital may be substituted for labor (and vice versa), the relationship between population growth and the level of per capita income is likely to be of modest importance overall except, perhaps, in the very long term. Moreover, the model predicts no relationship between the rates of growth of population and per capita output – at equilibrium, population is growing while per capita output is constant.

A second key component of the model – technological progress – offers the opportunity for a dynamic equilibrium of persistent growth in per capita output. Technological progress is characterized by an increase in output independent of an increase in inputs, and if it occurs at rate g, then in the long run output per capita will grow at rate g. Experience over the past century in industrialized economies has demonstrated the crucial role of technological progress in economic growth.

If this neoclassical model is correct, then barring a relationship between the rate of population growth and the rate of technological progress, and taking the saving rate as exogenous, there should be no correlation between the rate of population growth and the rate of per capita income growth in economies on their equilibrium growth paths. Indeed, when researchers compared the per capita economic growth rates of countries with their population growth rates, the data revealed no discernible relationship (US National Research Council, 1986; Blanchet, 1991).

This result has reinforced the agnosticism that neoclassical economists have generally expressed regarding rapid population growth and economic growth. The basic macroeconomic growth model gives no foundation for advising national policymakers that they would accelerate economic development by encouraging fertility decline. Nor does the conclusion necessarily change if rapid population growth is associated with a lower level (as opposed to growth rate) of per capita income. Neoclassical microeconomic theory is based on the assumption that there is a household-level trade-off between fertility and material standard of living, since both children and goods are sources of utility. A relationship between rapid population growth and lower per capita income at the macroeconomic level therefore may be consistent with the aggregation of utility-maximizing fertility decisions made at the household level.

However, more recent correlation studies (e.g., Kelley and Schmidt, 1996) have found a significant inverse relationship between the rates of growth of population and per capita income for the 1980s, although not for earlier years and not always for the postwar decades considered as a whole. According to an econometric argument (Blanchet, 1988), the observations are being generated by an underlying Malthusian model that correlation analysis is only now starting to reveal. Another explanation is that poor countries were displaced from their steady-state growth paths during the troubled decades of the 1970s and 1980s and, failing to adjust to external shocks, have remained displaced. This would be sufficient to lead to the emergence of a negative correlation (Lee, 1983, p. 54). Countries with rapidly expanding populations may also have found it more difficult to regain their steady-state paths than did countries with moderate rates of demographic increase. Complicating matters is the consideration that the current and lagged impacts of demographic increase may differ. A number of statistical analyses (Bloom and Freeman, 1988; Ahlburg, 1987) concluded that the rate of per capita economic growth is inversely correlated with the contemporaneous rate of population growth, but is positively correlated with the past (i.e., lagged) rate of population growth, mostly via labor supply effects.

Correlation studies, while influential in policy debate, are of limited relevance because the highly simplified models that underlie them may leave out a range of effects, many of them adverse, of high fertility and resulting rapid population growth. The most intensively investigated macroeconomic linkage has been that between savings and the rate of population growth, which we examine more closely in the next section. A rapidly growing, hence younger, population may exhibit a lower saving rate, which might retard economic development. One obvious extension of the model is to endogenize the saving rate as a function of the age structure of the population. Along the same lines, a population with many older persons drawing down accumulated assets in their retirement should save less than one in which many adult workers are saving for old age. Another possible extension is to endogenize the rate of population increase on the assumption that rates of fertility and mortality are a function of the level of per capita income. A third is to endogenize labor-force participation rates, which should, in theory, decline with rising per capita income as consumption of leisure increases. Finally, many authors have argued that technical progress is not exogenous, but may be accelerated by the pressure of population against fixed resources or may be impeded if demographic factors reduce investment in human capital.[2] Extensions such as these have been at the heart of the applied economic–demographic modeling literature.

3.1.3 Population growth in the basic model: A closer look

The basic model predicts that population growth has no effect on the equilibrium growth rate of per capita income, but reduces the level of per capita income through capital dilution. A number of other effects are possible as well. The simplest and most direct is an accounting effect on per capita income. Because a person born today does not become a productive worker for 15–25 years, population growth will increase the size of the population while temporarily leaving gross output constant. This "denominator effect" is relatively modest, however, and disappears in the long term.

Here we examine three less direct, but potentially more important, effects that have received considerable attention: the effect on physical capital formation through saving and investment, potentially positive effects on economic growth through economies of scale and the like, and the impact of population growth on the social sector.

Physical Capital Formation

As originally postulated by Coale and Hoover (1958), rapidly growing, and hence younger, populations are likely to save less than slowly growing older populations because a larger fraction of income must be spent on raising children. If true, then the capital dilution predicted by the basic model, resulting from faster-growing populations with a constant saving rate, would be exacerbated since the saving rate itself would decline. The lower saving rate would cause a sharper slowdown in capital accumulation, even lower productivity per worker, and depressed output. In contrast, if population growth is slowed, saving rates should increase as youth dependency declines, providing a boost to output per capita through capital deepening.

However, an enormous investment of research has left many issues still open. Over 20 years ago, Leibenstein (1976, p. 618) wrote that the issue of whether population growth reduced the availability of savings could not be answered based on the research available at that time. A decade later, McNicoll (1984, p. 206) wrote, "in general, rather little can be concluded," and suggested dryly that the phrase "based on presently available research" could be deleted without loss from Leibenstein's sentence. Research in the past 10 years has been generally favorable to refinements of the Coale–Hoover hypothesis (Mason, 1988; Higgins and Williamson, 1997), but many issues remain far from settled (Lee *et al.*, 1999).

Empirical studies have generally found only a weak statistical association between household youth dependency ratios and household savings (Hammer, 1985). The amount of savings is a function not only of the saving rate, but also of the level of income; thus, even if it is accepted that high fertility reduces the household saving rate, savings themselves might be unaffected, or even rise, if additional children contribute to household income either directly or indirectly by freeing adults for income-earning activities. A higher youth dependency ratio may encourage earlier labor-force entry or expanded labor supply on the part of other household members. Such behavior is likely to be particularly prevalent among the very poorest households, who may live so close to the subsistence level that the only possible strategy for coping with an increase in family size is a corresponding increase in production. Finally, even if it is accepted that high fertility reduces savings at the level of the individual household, compositional problems remain to be resolved at the level of the population. A rapidly growing population is one in which there are many young households saving for retirement relative to the number of elderly households drawing down accumulated assets – on compositional grounds, then, rapid population growth should increase the national saving rate (see Section 3.4).

It is possible that demographic increase might serve as an incentive for capital formation. By making labor more plentiful, population growth raises the return to capital relative to the real wage rate (Lee, 1980; Weir, 1991), thus favoring saving and investment. By temporarily threatening to reduce living standards (through the "denominator effect"), rapid population growth might even induce investment at both the national and household levels. Mortality decline should have an important accounting effect on investment decisions: by raising survival probabilities, reductions in mortality lengthen the horizon over which investment projects pay off. More important than accounting considerations may be the change in values that accompanies mortality decline: less fatalistic persons are more likely to perceive and exploit profitable investment opportunities.

On top of all these ambiguities surrounding the relationship between population growth and savings, empirical findings indicate that the contribution of physical capital formation to long-term economic expansion is less than had been believed. Kuznets (1966) and Denison (1985) were able to attribute only around 10% of American long-term economic growth to increases in the quantity of capital. More important were improvements in the quality of the labor force, economies of scale, agglomeration economies (i.e., economies that are made possible by the location of different economic activities in proximity to each other), technical change, and shifts of resources from low-productivity to high-productivity sectors. On the other hand, economic conditions in developing countries may imply a greater role for capital accumulation; a contribution of 25% may be reasonable (Hogendorn, 1990, pp. 91–92, for references).

Could Population Growth Accelerate Economic Growth?

In 1981, Julian Simon caused an uproar in the international population policy community with the publication of *The Ultimate Resource*, a book which argued that, far from causing stagnation, rapid population growth might accelerate economic progress (Simon, 1981). Somewhat lost in the controversy was the fact that Simon was simply putting forth, albeit in a rather aggressive tone, points that had long been standard in the economics of population. Moreover, in the years since publication of *The Ultimate Resource*, economic research has focused increasingly on these positive impacts of population growth.

Public goods and infrastructure.   Public goods, such as agricultural research and development institutions and rural extension services, are goods for which indivisibilities in consumption exist; that is, the amount consumed by person A does not diminish the amount consumable by person B. Infrastructure, while it may be congestible, usually has significant public-good aspects. When supply of a public good is not fixed, it is cheaper on a per capita basis to produce the public good in a large population than in a small one. A similar point can be made in the case of technological change and innovation, which we discuss below. The larger the labor force, the more broadly the fixed costs of research and development and of adopting new technologies can be spread.

Economies of scale.   At the level of the firm, economies of scale exist if a given increase in all inputs leads to a more-than-proportional increase in output. At the industry level, economies of scale tend to be exhausted at moderate market size; moreover, it is not the size of the local market but the size of the local and export markets together that is of importance. Agglomeration and urbanization economies are cost advantages that arise from the spatial concentration of production; in most sectors, these can be fully reaped once a country has one city of moderate size. On both these counts, economies of scale arising from rapid population growth are rather unlikely to be a source of economic growth in most less developed countries (LDCs).

"Vintage" arguments.   The embodiment of technological change in capital or, due to improving standards of training and education, in workers has long been known (e.g., Leibenstein, 1967) to favor rapid population growth via "vintage" logic. A more rapidly growing population will have a younger and hence, despite its inexperience, more innovative and aggressive labor force; a more rapidly expanding capital stock will consist, on average, of newer machines. The benefits from technical advancement embodied in the labor force have been mathematically investigated by van Imroff (1988), who added to a neoclassical growth model the assumption that only new workers use new technologies.

Technological change and innovation.   In current economic growth theory, often called "endogenous growth" theory, technical progress and the rate at which new technologies are adopted are endogenous variables (Romer, 1990). Mathematical models in which density-dependent technological change allows escape from Malthusian constraints have been elaborated by Steinmann and Komlos (1988) and Komlos and Artzrouni (1990).   Simon (1976) made an ad hoc attempt to make this theme operational by replacing the constant term in a Cobb–Douglas production function with an exponential term whose rate of growth depended on the rate of growth of the labor force.[3]   But the forces that lead to endogenous growth – learning by doing and cumulative returns to scale – are found in knowledge-intensive industries. If rapid population growth were to concentrate economic activity in low-technology sectors (like traditional agriculture), its effect would be to retard rather than accelerate economic growth.[4]   As discussed further in Chapter 5, rapid population growth might give rise to induced innovation within agriculture, but if agriculture is an inherently low-productivity sector, macroeconomic growth would suffer nonetheless. Across-the-board productivity gain induced by population pressure is not substantiated as a general proposition, especially in the near and medium terms. As the time horizon lengthens, however, it becomes increasingly unrealistic to assume away the possibility of endogenous technical progress related to population growth.

Does Rapid Population Growth Overburden the Social Sectors?

One component of poverty is inadequate access to education and health care. Rapid population growth and a youthful age structure strain public resources in both areas (Najafizadah and Mennerick, 1988, and Jones, 1990, for education; Jones, 1990, for health). It is difficult in the case of either sector, however, to assess the resulting damage. Schultz (1987) found that rapid population growth did not reduce the quantity of education provided (i.e., enrollment rates), but reduced its quality (i.e., teacher–student ratios, availability of textbooks, etc.). Kelley (1996) found a significant inverse relationship between rates of population growth and secondary school enrollment rates, but no relationship between number of school-age children and educational attainment. Whatever the effect of demographic dilution of educational resources, it is secondary compared with inefficiencies arising from the highly regressive subsidy of secondary and university education at the expense of primary schooling (Jiminez, 1989), as well as the inequitable distribution of resources between boys and girls. Likewise, the overburdening of health resources is less serious, and certainly less amenable to policy intervention, than misallocation of resources toward curative hospital care at the expense of primary care (World Bank, 1993).

On the other hand, Birdsall (1994b) points out that it can be more efficient to address strains on educational systems through measures to reduce family size (thereby increasing per child family expenditures on education) than through subsidies to education. One reason is the inherent administrative difficulty of designing efficient subsidies; another is the possibility that by lowering the costs of children without changing benefits, subsidies may encourage higher fertility, placing a further burden on the educational system.

3.2 Population and Environment

3.2.1 A neoclassical model of population, economic development, and environment

Neoclassical economic growth models implicitly assume an underlying household-level trade-off between number of children (as well as so-called child quality) and consumption of all other goods. When extended to include the environment, such models also assume trade-offs between the quality of the environment and both number of children and material consumption. Trade-offs are understood in terms of marginal economic costs and benefits (Kneese, 1988; Cropper and Oates, 1992); nature is valued in instrumental terms (Oksanen, 1997), that is, only to the extent that it is a source of utility to individuals. If, for example, individuals derive more utility from a highly developed national park than from an unspoiled wilderness area, it follows that the wilderness should be developed.

Natural resource supply and demand are mediated through markets, with prices providing the scarcity signals that stimulate substitution (which is always assumed to be an option), exploration, conservation, and (in the case of the environment) cleanup.[5] For example, as copper becomes scarce and its price rises relative to prices of other materials, producers turn to alternatives in manufacturing, investors allocate funds to exploring for and developing new copper sources (thereby profiting from the rising price), recycling becomes attractive, consumers substitute away from goods containing a great deal of copper, and so on. Shifting prices also incite technical progress by alerting the research and development community to profitable avenues of technology development.

The cornerstone of neoclassical natural resource and environmental economics is that all natural resource prices contain an embedded "scarcity rent" – that is, a return on the resource due to the fact that its supply is limited by nature. This scarcity rent, the asset value of a unit of the resource in situ – a tree in the forest, say, or a barrel of oil in the ground – should increase exponentially over time at the financial rate of interest. If this were not the case, an owner would liquidate the resource stock immediately and invest the proceeds in financial assets.

By holding forth the promise of a higher price in the future, scarcity rent encourages natural resource owners to conserve supplies for the future.[6] Many instances of wasteful overexploitation of natural resources can be explained by the fact that no single decision maker is able to earn the scarcity rent; this is the neoclassical framing of the "tragedy of the commons" (see discussion below).

Partly because the share of primary goods (especially food) in the consumption basket falls as consumer income rises, the share of natural resources in economic output as a whole declines as a consequence of development. Economic growth need not place pressure on the resource base, because it generates a rising surplus (in the form of output and income outside the primary sector) that can be invested in the primary sector so as to ensure continued supply of natural resources. The quality of the environment should rise along with per capita income: demand for a clean environment goes up, and along with it, availability of the means to produce it. Demand for environmental quality includes "existence demand" and "options demand," for example, preservation of species because people place value both on

---

Box 3.1. Time discounting.

Discounting allows the comparison of economic costs and benefits occurring at different times; the higher the discount rate, the less the well-being of future generations is valued relative to that of the present generation. Two general approaches to setting the social discount rate have been used (Arrow *et al.*, 1996). The first is a prescriptive approach, which seeks to determine an appropriate discount rate based primarily on ethical principles. Prescriptive discount rates are composed of two components. The first is based on the decreasing marginal utility of consumption and the expectation of rising consumption levels in the future; that is, a dollar's worth of cost or benefit 50 years from now should count less than a dollar's worth today, because people will be better off in the future and will therefore derive less utility from a marginal dollar. The second component, called the pure rate of time preference, reflects the societal preference for consumption now versus consumption in the future. The prescriptive approach finds little ethical basis for a nonzero pure rate of time preference, and therefore generally yields discount rates of at most a few percent per year based mainly on decreasing marginal utility of consumption.

The second approach to discounting is descriptive and attempts to derive the discount rate based on observations of actual behavior. It generally arrives at higher rates because it includes positive rates of pure time preference derived from empirical data on the opportunity cost of capital. That is, since investments in mitigating climate change, for example, could have been invested in other activities producing a positive rate of return, benefits of mitigation projects must be discounted at this rate to ensure the most productive use of capital. While there is general agreement that some form of discounting is required in cost–benefit analysis, there is little agreement on the best approach (Schelling, 1995; Rabl, 1996).

---

knowing that species exist and on preserving them in case they are of economic value in the future.

Neoclassical "sustainability" is defined in terms of intergenerational transfers: natural resource consumption decisions in the present period are sustainable if the natural environment which is bequeathed to the next generation is not inferior in value to that which was inherited by the present generation.[7] The key phrase is "in value," which opens the door to the expression of individual preferences via the marketplace and which, in particular, admits time discounting (see *Box 3.1*) to reflect preference for consumption now instead of later (Pearce *et al.*, 1990).

3.2.2   Population growth in the basic model including the environment

Abstracting from associated changes in age structure (which may shift the structure of demand), and assuming there is no market failure, the effect of population growth is to drive up scarcity rent and thereby raise the price of natural resources relative to the price of labor. As a result, wages decline while income rises for resource owners.[8]   Since resource owners disproportionately belong to high-income groups, rapid population growth can be expected to raise income inequality and increase the number of poor people as a fraction of the population. As we discuss below, these distributional consequences provide one basis for population policies, but the empirical evidence is not strong.

Another basis for policy is that population growth can exacerbate or give rise to market failures involving natural resources and the environment. Among the various forms of market failure, most attention has focused on the linkage between population and common property problems, that is, the "tragedy of the commons."

The Tragedy of the Commons

The tragedy of the commons has been a powerful metaphor since the publication of Hardin's (1968) much-cited article of that title. It describes the degradation of a shared resource under conditions of uncontrolled access. Framed within a parable of a village grazing commons, degradation is a direct result of a system of incentives that encourages villagers to graze additional cattle. A villager reaps the full benefit of each additional animal's grazing, while the whole village shares the cost of the reduction in available grass. As each villager pursues his or her own personal gain, the number of cattle rises until it eventually exceeds the carrying capacity of the commons.

According to Hardin, population growth represents a tragedy of the commons. However, a neoclassical reading of the situation would conclude that the key to the problem is not population per se, nor is it common ownership, rather, it is the failure to control access to the commons. Many resources that are held in common (communal grazing lands in semiarid regions, for example) are utilized only according to strictly enforced social rules (Orstom, 1990). In principle, controlling access through the imposition of user fees, privatization, legislation of access rights, or any other appropriate method could achieve an optimal level of commons usage for any given population size. In fact, in some cases population pressure may even encourage privatization of formerly common property resources and evolution of rules limiting access to resources still held in common (see Hayami and Ruttan, 1991, for an example from the Philippines). However, in Côte d'Ivoire, Ahuja (1998) concluded that rapid population growth, by raising monitoring and enforcement costs, significantly weakened the ability of communities to manage communal lands.

> **Box 3.2.** Externalities to childbearing.
>
> The most comprehensive empirical work in this area has been done by Lee (1989, 1990) and Lee and Miller (1991), who estimated nonenvironmental reproductive externalities in a number of countries in the 1980s that derived from three main sources. The first source, *collective wealth*, consists of assets to which all members of society have free access and which are diminished in value, on a per capita basis, as population grows. Examples include common property resources such as publicly owned land, state industries, and national rights to fisheries. Second, the provision of *public goods* is also susceptible to externalities. Because these goods are generally cheaper to produce in larger populations, population growth gives rise to a positive externality. Third, externalities to childbearing also arise from *intergenerational transfers*. Different age groups within a population will have different balances of public services received and taxes paid; in general, young and old age groups are net recipients of transfers for health, education, and pensions, while the middle-aged are net sources of transfers.
>
> Cautioning that their estimates are rough and do not include environmental effects (such as climate change), Lee and Miller (1991) nonetheless conclude that net externalities to childbearing are positive in more developed countries (MDCs) and negative in less developed countries (LDCs). The positive external effect of an MDC birth arises mostly from the contributions an extra member of the population will make supporting transfers to the elderly. The negative external effect of an LDC birth results from dilution of collective wealth and is significant only in countries with substantial endowments of natural resources such as oil, coal, or forests.

The Second Tragedy of the Commons

Lee (1991) has argued that high fertility gives rise to a "second tragedy of the commons." Say that a user fee is imposed on those desiring access to the resource, and that this user fee is raised to reflect the increase in scarcity that is consequent upon a marginal birth. The resource is optimally managed before and after the birth, and neither situation is more desirable than the other on economic efficiency grounds. But the higher user fee will be paid, not only by the parents who decided to have the child, but by all families. In this sense, there is an "externality" associated with fertility that cannot be internalized by any policy impinging only on the resource, but must be addressed directly by policies affecting childbearing (see *Box 3.2*). Externalities are defined as costs and benefits arising from production and consumption decisions that are not borne by the decision maker. In the presence of negative externalities, the free market will allocate too much labor and capital to production of the good in question, and in the case of external benefits, too little.

Estimates in the context of climate change have generally concluded that the negative greenhouse externality to childbearing is large (see *Box 6.3*). That is, each additional birth imposes additional costs on society in order to offset the extra GHG emissions resulting from the marginal increase in population. We take up this issue and its relevance to policy in greater detail in the final chapter of the book.

3.2.3 Ecological economics

In the model of population, economic development, and environment described above, the main mechanism of change is substitution in response to price signals. Population size is important mostly because it shifts relative prices. However, a number of other perspectives exist (Jolly, 1993). Ecological economics, for example, views population size as a crucial (and dangerous) scale factor. If absolute ecological limits are close enough to be relevant, as some argue, then ecological economics, as propounded by Herman Daly (1977) and others, is a more appropriate framework than neoclassical economics.

Ecological economics combines elements of economics and ecology with concern for distributional issues. When the scale of total human activity – population times per capita output or impact – was tiny relative to the size of the ecosystem, economic growth passed for social progress and appeared to offer a panacea for social ills. To ecological economists, however, the signs are multiplying that ecological limits have been exceeded (Goodland, 1992). Under such circumstances, total throughput of the natural system must be stabilized, or even better, reduced. Because further growth is ecologically reckless, social progress can be achieved only by enhancing equity. Since equity is an explicitly normative concept, ecological economics is a self-professed "post-normal" science (Duchin, 1994; Funtowicz and Ravetz, 1994; M'Gonigle, 1999), driven by values and combining advocacy with analysis.

Scale, and hence aggregate population size, is virtually irrelevant in neoclassical economics; it is central to ecological economics (Daly, 1992). The disagreement between the two perspectives may be ascribed to fundamentally different pre-analytic visions of scarcity. Scarcity in neoclassical economics is relative (scarcity of good A relative to good B), which drives allocative decisions (for example, allocation of capital and labor between sectors A and B), which in turn balance out relative scarcities. Although the allocative focus implicitly recognizes that each individual human activity has an optimum scale (allocation of productive resources to agriculture versus industry, for example), there is no such thing as an overall optimum scale of human activity. By extension, there is no such thing as an optimum size for the human population.

The pre-analytic vision of scarcity in ecological economics is not relative, but generalized, scarcity (e.g., Daly, 1977, 1991) driven by growing scarcity of energy.[9] Under conditions of relative scarcity, the potential to substitute abundant materials for scarce ones is practically infinite (Goeller and Weinberg, 1976; Goeller and Zucker, 1984). Under conditions of generalized scarcity, the price of virtually all goods that humans consume would rise relative to the price of the one service they produce, labor. Forced to allocate an ever larger share of productive resources to eking out food and materials from ever costlier sources (and cleaning up pollution), society would find itself caught in a vicious circle. Population growth worsens generalized scarcity, thus ecological economists, in contrast to neoclassical economists, are deeply concerned by current population trends.

Fundamental to the view of population in ecological economics is the concept of carrying capacity. The carrying capacity of an ecosystem is the largest population that can be supported without diminishing the ecosystem's supportive capacity. Cohen (1995) has published an encyclopedic review of estimates of the human carrying capacity of the Earth made since the 17th century. With a few exceptions, these cluster in the range of 4 to 16 billion – by coincidence, a range very close to the range of population projections discussed in Chapter 2. Cohen (1995) has argued that a family of human carrying capacity surfaces – maximum supportable population as a function of efficiency in natural resource utilization and level of consumption per capita – might be a useful indicator of the large-scale consequences of individual preferences and values.[10] Yet social scientists – especially, but not only, economists – have shown little interest in applying the carrying capacity concept to humans. One obvious reason is that culture, institutions, accumulation, etc., allow humans to change the technical coefficients that must be assumed in order to calculate carrying capacity. The elements of such a model of population and the environment were enumerated by Duncan (1959, 1961) in the POET model: the human population (P), socioeconomic organization (O), the environment (E), and technology (T). Early human ecologists treated the natural environment, social organization, and technology as exogenous factors to which the human population adapted. Over time, however, the tendency has increasingly been to view population as an active element of the system. Today, population is viewed as acting not only on the environment, but also on technology (Boserup, 1965, 1981) and social structure and organizations (McNicoll, 1990, 1993).

So which view, neoclassical economics or ecological economics, is correct? Disregarding discontinuities in the form of "surprises," it should be possible to judge the relative merits of the two schools of thought on the basis of time trends in scarcity indices, such as per capita resource production, relative natural resource prices, real unit extraction costs, and so on. Some 10 years ago, economists and environmental advocates fought pitched battles on the basis of such data and, largely as a result of injuries inflicted on the credibility of the "limits to growth" model during the course of this debate, US policy in the area of global sustainability turned sharply toward laissez-faire (MacKellar and Vining, 1988). One comprehensive review of scarcity index trends reached relatively pessimistic conclusions in the area of energy, relatively optimistic conclusions in the area of agriculture, and mixed conclusions in the area of forestry products (MacKellar and Vining, 1987). On the other hand, the authors themselves described the shortcomings of various scarcity indices as "practically irremediable" (MacKellar and Vining, 1989). Using energy costs of resource extraction (including "net energy gain" in the extraction of energy) as their index of scarcity, ecological economists have found that scarcity has been increasing (Hall *et al.*, 1986; Cleveland, 1991). On the other hand, the natural resource scarcity debate appears in some degree to have cooled down, in part because concern has shifted away from depletable resources and toward renewable resources and environmental quality, where comparable time series data are difficult to find.

3.3   Vicious-Circle Models

One extension of the basic neoclassical economic model of population and the environment encompasses poverty and the low status of women and children; in so doing it adds a poverty trap based not on the macroeconomic reasoning of Malthus, but on microeconomic effects at the household and community levels (World Bank, 1992). This model is motivated by the belief that high fertility, poverty, low status of women and children, and environmental damage are bound up in a web of interactions (see *Figure 3.1*) in such a way that stresses from any one of these sources can trap societies in a vicious circle of destructive responses (Markandya, 1998).

The case in theory for a link between poverty and environmental degradation is strong. Impoverished, insecure households living at the margin of survival operate with a high discount rate (Holden *et al.*, 1998) and are therefore likely to deplete surrounding natural resources and environmental assets with little heed for future consequences.[11] Even if such households desired to safeguard their surroundings, however, their room to maneuver – in the form of the types of substitution, adaptation, and behavioral change that are at the heart of neoclassical economics – would be restricted. For example, when poor households intensify agricultural production to cope with rapid population growth, they are unable to afford the investment and inputs necessary to conserve soil. Thus, poverty leads to environmental degradation by impeding substitution and adaptation responses to ecological stress; environmental degradation leads to poverty since the renewable natural resource base is a key asset of populations at risk of poverty.

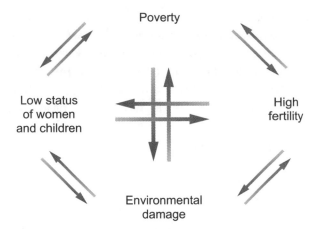

Figure 3.1. A vicious-circle model.

Dasgupta (1993, 1995) adds high fertility and gender inequality to the picture. Rapid population growth places pressure on the renewable resource base – say the supply of wood fuel.[12] Poverty prevents substitution of alternative fuel sources; the low status of women and girls devalues the increasing time and effort that they must devote to daily gathering of wood and other environmental products (Cooke, 1998; Agarwal, 1994; Sen, 1994a). As the value of child labor rises, fertility decline is impeded both directly (the economic value of children rises; e.g., Feldman, 1990) and indirectly (attitudinal change that might occur through education of girls is blocked because girls are kept home to help their mothers). The result is faster population growth, further degradation of the renewable resource base, and yet further erosion of the position of women. Poverty, depletion of renewable natural resources, high fertility, and low status of women combine in a vicious circle which forms a poverty trap and may lead to local ecological breakdown.

Vicious-circle reasoning represents an important contribution to theory: it points to a single unified theory of fertility, poverty, and environmental degradation. It also places equity concerns, in the form of poverty and unfair treatment of women, at the center of the policy picture. Because the feedback effects in the model sketched above are so dense, any policy that improves the status of women, alleviates environmental stress, or reduces poverty is bound to have positive repercussions throughout the system. On the other hand, two notes of caution are in order.

Gender and Equity Concerns

While an equity-based view will find much to agree with in this perspective, it is bound to treat other aspects with suspicion. Jackson (1995), far from seeing vicious-circle reasoning as serving women's interests, sees it as a means of co-opting women's legitimate interests in a way congenial to the established policy orthodoxy. It is, for example, indicative that women's status and girls' education are treated as nearly equivalent despite the fact that women's status is a complex, multidimensional variable that cannot be adequately represented by a single proxy variable (Balk, 1994). Education of girls may get a disproportionate share of attention in part because it can so easily be translated into targets and indicators – female literacy rate, female enrollment rates, etc.[13]

Global Ecological Concerns

According to some ecologists, regardless of how population growth interacts with poverty, it places the global ecosystem at risk of catastrophe. Vicious-circle reasoning, however, is oriented toward the welfare of the individual. It takes little stock of the concerns raised by the complex dynamics that underlie modern ecology (Holling, 1994), according to which seemingly small alterations in the present period and at the local scale can have dramatic, unforeseeable consequences in later periods and at the global scale. The World Bank estimates that one-third of the population of LDCs lives in poverty, of which one-half lives in environmentally fragile zones. Vicious-circle logic, therefore, only applies to one-sixth of the LDC population, less than one-seventh of the total world population. Policies to encourage fertility decline among these subpopulations are of limited impact in terms of stabilizing world population growth. Derived almost exclusively in a rural setting, vicious-circle reasoning will have to be refocused and extended to be relevant to the urban population (see *Box 2.1*). Finally, the time scale must be considered. Even at the local level, interventions to address problems such as open access to resources, high fertility, poverty, etc., are likely to have an impact only in the longer term (Bluffstone, 1998).

Empirical evidence

Empirical evidence on determinants of fertility is discussed in Section 2.2.2. In this section, we summarize evidence on the impacts of high fertility pictured in *Figure 3.1*.

Does High Fertility Cause Environmental Damage?

Although it is almost entirely unsystematic and sometimes anecdotal, evidence is abundant (United Nations Secretariat, 1991; Panayotou, 1996) that rapid population growth exacerbates a range of renewable resource problems, including tropical deforestation, soil erosion, overgrazing of semiarid lands, and overfishing of coastal waters. However, the main theme that emerges from research is that, while population pressure may appear as a proximate cause of environmental damage, its effect is mediated through factors such as poverty, market failure, inadequate access to resources necessary for investment, inadequate and inappropriate institutions, and misguided macroeconomic policies. One attempt to design a comprehensive conceptual framework for analyzing population and land degradation (Bilsborrow, 1987) identified no fewer than nine contextual variables that determined the impact of population pressure.[14] The formula by which high fertility is translated into environmental damage includes inequitable access to resources of all kinds (including land, credit to finance investments, and insurance to cover risks), insecurity, poor governance, and lack of opportunities for alternative economic activities or out-migration. At the economy-wide level, a theme which emerges is that it is not growth in population itself which gives rise to environmental costs, but rather growth in labor force relative to growth in human and physical capital (Abler *et al.*, 1998). In conclusion, high fertility can cause environmental damage, but whether it does or not is dependent on the nature of markets and many other institutions.

Does High Fertility Cause Low Child Quality?

There is strong evidence that high fertility and large family size are inversely correlated with children's living conditions and education and, as a result, may eventually give rise to low levels of human capital (MacKellar, 1994, for citations). Broadly defined, human capital includes not only training and education, but also health, literacy, adequacy of nutrition, etc. The impact of family size on child welfare may be attenuated in societies where extensive community resources and extended family resources are invested in child rearing (Desai, 1995).

Rapid population growth is generally associated with short birth intervals, and many studies have concluded that closely spaced children are subject to higher mortality (Hobcroft *et al.*, 1985; Potter, 1991, pp. 224–230, for references). In the Matlab district of Bangladesh, this elevated mortality is concentrated in the 2–4-year-old age group (Le Grand and Phillips, 1996). The impact of close birth spacing appears to be especially deleterious for girls (Muhuri and Menken, 1997). This would strongly suggest that those who survive under such circumstances suffer from poorer health than more widely spaced children. Cross-household studies have found that the number of children in the household is inversely correlated with child nutrition and educational expenditure per child. There is also evidence that unwanted births, which are likely to be higher-parity births, are associated with lower school enrollment rates for all children in the family.

Inadequate birth spacing also adversely affects maternal health, and the demands of pregnancy, lactation, and child care associated with high fertility reduce the rate of return to education of women. In view of the very strong correlation observed between mother's education and child survival (Cleland and van Ginnekan, 1988; Cleland, 1990), this may perpetuate the destructive cycle of unhealthy mothers and unhealthy children.

Montgomery and Lloyd (1996) argue that such correlations provide "reasonably reliable guides" to underlying causal relations and, by implication, for the design of policy interventions. On the other hand, there can be little doubt that correlation studies overestimate the strength of causal effects.

High fertility and child welfare can be causally linked in three ways:

1. Low standard of living may cause both high fertility and poor living conditions.
2. High fertility may worsen children's living conditions holding income constant (for example, by reducing parental attention per child).
3. High fertility may reduce endowments of nutrition and education per child.

Rosenzweig (1988, 1990) examined causality in the area of high fertility and child welfare and wrote (1988, p. 83), "each assertion [points 1–3 above] can be supported. But the quantitative evidence for any one is not overwhelming; there are only a few studies in each case which go beyond correlations, and the estimated magnitudes of the causal relationships are small."

For example, studies from India, Indonesia, and Malaysia indicate that, on average, the unanticipated extra birth due to a twin would reduce the school enrollment rates of other siblings by one-third if the birth occurred at the third or fourth pregnancy, and rather less if earlier. A doubling of birth intervals was estimated to raise birth weight by 3–6%. Such estimates are significant, but modest, and their impact would be conditional on the overall level of socioeconomic development, that is, the quality of available education and the prevailing level of nutrition. More recent studies have continued to indicate that caution is warranted. In Vietnam, Ahn *et al.* (1998) found that, once intervening variables were taken into account, the strong simple correlation between family size and children's education was greatly weakened. Desai and Alva (1998) found that much of the observed correlation between mothers' education and child health in Vietnam was due to the fact that both variables are positively associated with income. However, also in Vietnam, Houghton and Houghton (1997) found an inverse relationship between birth order and child nutrition.

In conclusion, evidence shows that high fertility does have a negative impact on child welfare. However, the magnitude of the impact is modest, and certainly much lower than would be concluded on the basis of correlation studies.

Does High Fertility Lower the Status of Women?

In addition to diluting household resources available to all children, high fertility has three effects that fall disproportionately on girls (Lloyd, 1994):

- An *opportunity effect*, whereby high fertility may lead parents to deny girls access to available resources (such as schooling).
- An *equity effect*, whereby high fertility may make more unequal the distribution of available resources among household members.
- An *intergenerational effect*, whereby high fertility may impede attitudinal change.

The first effect operates through lower income or poverty; as just discussed, the link between high fertility and this variable is probably modest at the household level. However, there is evidence that when impoverishment strikes, girls' access to education suffers more than boys'.

Does High Fertility Cause Poverty?

Poverty is usually, although not always, positively correlated with household size (e.g., House, 1987). However, major surveys have failed to strongly endorse the hypothesis that high fertility causes poverty. According to Rodgers (1984, p. 169), high fertility appears to play some causal role, but the effect is weak. Birdsall and Griffin (1988, p. 49) found that poverty encouraged large family size but concluded that, regarding the link in the opposite direction, there was "some theory but much less evidence." Ahlburg (1996) found the evidence that high fertility and rapid population growth cause poverty to be far from clear.

In all matters concerning poverty, the labor market plays a crucial role. Demographic increase has both behavioral and accounting effects on labor supply. Labor supply behavior is so inextricably tied up with labor market structure (Rodgers, 1989) that it is difficult to generalize across different settings about the effect of demographic factors. Two stylized facts which have long been known are, first, that high fertility is correlated with low rates of female labor-force participation in the modern sector and high rates of female labor-force participation in the informal sector (e.g., Standing, 1978); and second, that children make substantial contributions to home production (e.g., Mueller, 1976). The latter might, as one of its effects, free adults for market-based production; however, some studies have revealed that, in the case of women and girls, large household size may merely add to the intramural workload with no compensating reduction in extramural labor supply (Lloyd, 1994, for references).

The accounting effect is straightforward: a rapidly expanding labor force will be a relatively young and inexperienced one. Rapid growth in the labor force should reduce the average wage rate relative to the price of capital and thereby promote the adoption of labor-intensive modes of production. Research by Boserup (1965, 1981) has amply illustrated this process in the case of agriculture. In industry, however, especially in the formal sector, the scope for substitution in production may be quite limited (Egea, 1990). At the national level, studies have revealed no correlation between the rate of population growth and the rate of open unemployment, but the latter is a notoriously inadequate index of labor market activities. The most important study of labor markets and employment in LDCs is still that of Bloom and Freeman (1987), who found scattered empirical support for the proposition that rapid population growth lowers wages, but reached no general conclusion. Many LDCs have succeeded in absorbing large shares of their rapidly growing populations into productive employment, but this is no ground for optimism if wages have declined and workers have been forced into marginal activities.

If formal sector wages are rigid, then rapid population growth will increase the proportion of workers in the informal sector, placing downward pressure on informal sector earnings and widening the formal–informal sector income gap. An identical effect will operate within the informal sector itself, where entry into more desirable occupations is blocked by capital and skill requirements (Fields, 1990; Cole and Fayissa, 1991; House, 1992). Rapid population growth thus pushes workers down the productivity ladder, concentrating the least well endowed (in terms of human capital) in marginal activities. Empirical evidence is spotty, but something along these lines is suggested by Behrman and Birdsall's (1988) finding that the negative correlation between cohort size and wages in Brazil was much stronger for unskilled workers than for skilled workers.

In the countryside, the poverty effects of high fertility are linked with the markets for land and agricultural labor. Rapid population growth reduces the size of the average landholding, with the most adverse effect being felt by the poorest landholders, who resort to uneconomic subdivision and fragmentation and are thereby placed at increased risk of landlessness (Chaudhury, 1989, for the case of Bangladesh). Landlessness is, in turn, statistically associated with poverty. However, a survey of Asian countries (Khan, 1988) found that this stylized sequence of events was the exception rather than the rule. Among the factors that intervened between demographic increase and landlessness were expansion of cropland, increased labor absorption on existing land resources, industrialization, and growth of nonfarm employment. In India, Crook (1996) found that landlessness and agricultural productivity, not the rate of population growth, were the best predictors of state-level poverty rates.

Rural population growth may nonetheless reduce rural incomes. Evenson (1988) estimated for Northern India that, excluding Boserup–Simon effects (see endnote [3]), the elasticity of the per capita income of landless farm laborers with respect to population density was –1.5 and that of landlords' income was +2.5, making for a population-wide income elasticity of –0.5.

In conclusion, theory and limited evidence are far removed from the Malthusian image of poverty following ineluctably on the heels of population growth. Operating through the labor market and the market for agricultural land, rapid population growth might push some families on the edge of poverty over the line. It seems unlikely, however, that it would give rise to impoverishment on a large scale.

3.4 Economic Impacts of Population Aging

Research on population and the environment has focused on the impact of population size and rate of growth, not structure. However, the aging of populations and the labor force, especially in Europe, North America, and Japan, is a major emerging source of policy concern (OECD, 1998). Aging is also emerging as a major issue in LDCs (World Bank, 1994), not to mention in the formerly socialist countries of Eastern Europe and the former Soviet Union, which have experienced radical fertility declines in recent years. Population aging will have pervasive social and economic effects, some of which may ease pressures on the environment and some of which may worsen them. The fiscal and economic impacts of aging, while imperfectly understood, have been the subject of far more research than its links with the environment. It is broadly agreed that, just as individual decisions resulting in high fertility may give rise to external environmental costs, individual decisions resulting in low or sub-replacement fertility may impose external costs (on future generations) in terms of fiscal and other economic impacts of population aging.

One way of looking at population aging is in terms of possible impacts on productivity. A plausible fear is that the aging of the work force replaces young, dynamic workers with old, less productive ones, thus acting as a brake on progress. This microeconomic "productivity pessimism" has, however, received little support from research (Disney, 1996; Jackson, 1998). What older workers lack in dynamism, they may make up for in terms of experience. In addition, they have been sorted and matched with the tasks they perform best, a process which is still ongoing for young workers. The observed reluctance of firms to hire or train older workers has more to do with institutional barriers and inefficiencies than it does with low productivity of older workers. Other arguments relating population aging to productivity decline are macroeconomic in nature. Some of these, having to do with age of the capital stock and economies of scale in provision of public goods and the development and adoption of new technologies, were mentioned in Section 3.1.3. While plausible, all these arguments are rather speculative in nature. Equally appealing is the argument, which finds some empirical support (Cutler *et al.*, 1990), that technical progress will be accelerated as population aging makes labor more scare (Habakkuk, 1962).

Since productivity effects are difficult to predict, economic analysis of population aging has focused more on impacts on savings and capital formation. There are only three ways to finance consumption by the elderly. One strategy is for current workers to forgo consumption in order to transfer income to current retirees, either directly at the family level or indirectly through a balanced pay as you go (PAYG) public pension system. Assuming that the elderly consume all their income (i.e., transfers) and that workers save at least some of theirs, then the impact of population aging will be to reduce the aggregate saving rate [net savings relative to gross domestic product (GDP) minus depreciation] on compositional grounds.[15] Another strategy is for the state to borrow in order to finance consumption by the elderly through the public pension system; in this case, population aging will increase the public-sector deficit and, again, reduce the aggregate saving rate. The financing of pension system deficits through general revenue, as in some Western European countries, or the use of pension system surpluses to relax fiscal constraints elsewhere in the government budget, as many believe to be the case in the United States, are cases in point. The only remaining strategy is for today's workers to save (whether as individuals or through a funded public pension system) in order to finance their own retirement. In this case, current population age-structure trends in MDCs will act to increase the saving rate as baby boomers move through their peak saving years. Starting about 2010, however, retiring baby boomers will begin to sell off the assets they have accumulated (to younger persons who are accumulating assets for their own retirement) and convert the proceeds into consumption. Thus, even when individuals accumulate savings for their old age, the long-run result of aging is still, on compositional grounds, a decline in the saving rate. The advantage of an accumulation-based strategy is that higher saving in the near term will translate into higher capital formation, which will result in higher aggregate output in the long run. The debate over how best to finance the growing consumption needs of the elderly is, however, far from simple. A recent Organisation for Economic Co-operation and Development (OECD) survey concluded that while tax-favored private retirement saving accounts tend on balance to raise national saving rates, the impact is modest once the fiscal impacts of lower government tax revenues are taken into account (Kohl and O' Brian, 1998).

A number of empirical analyses have estimated the impact of projected population aging on private savings (i.e., household and corporate) and public savings (i.e., the public sector balance) under assumptions which amount to an extrapolation of current trends. These "business as usual" or "no response" results are, without exception, dramatically pessimistic. Roseveare *et al.* (1996) examined public pension and health care spending programs in 20 OECD countries. They estimated that for the OECD as a whole, the aggregate (public plus private) saving rate could decline from 7.4% in 1995 to almost zero, or even turn negative, by 2030. Franco and Munzi (1997) performed a similar analysis of the members of the European Union, reaching similarly pessimistic conclusions. Scheiber and Shoven (1994) projected that the US private pension system, which has been a large net source of savings to the economy because of the surplus of contributions over benefits and strong investment income, will be a net dissaver after 2025. In other words, pension funds will pay out more in benefits (which finance consumption) than they will take in as contributions and investment income.

Other compositional effects, however, are operating in addition to the increase in the share of the population in the retirement age bracket. Rapid economic growth raises the income of the young relative to the income of the old, which, if the young have higher saving rates in accordance with the life-cycle hypothesis (LCH), should increase the private saving rate. Slow economic growth, which has characterized the MDCs since the 1970s due to slower productivity growth, has the opposite effect There may, moreover, be cohort effects, with different birth cohorts having different propensities to save at the same age. Accounting for these three effects (change in the age distribution of the population, change in the age distribution of income, and cohort effects), Börsch-Supan (1991) projected OECD private savings from 1990 through 2020. Combining all three effects, he found that the total supply of private savings should continue to grow throughout the period, in large part because the baby boom cohort is characterized by unusually high saving rates. However, assuming constant tax rates, this increase in the supply of private savings will be more than wiped out in the early years of the 21st century by the reduction in public sector savings caused by rising age-related expenditure (essentially, pensions and health).

The studies above have concentrated on MDCs, but demographic aging is also on the horizon in a number of LDCs. In fact, a majority of the world's elderly population already lives in LDCs. While they still have time, national policymakers in this part of the world need to consider carefully what public and private institutions will be best suited to provide the income and health care needs of the elderly (World Bank, 1994). From the point of view of the global macroeconomy, the role of Asia is especially important for three reasons. First, the so-called Asian "Tigers" are aging fast because of the rapidity of fertility decline in the region. China, because of the one-child policy, represents an extreme case. Second, because economic growth in the region has been more rapid than in the rest of the world, the Tigers account for a rising share of world income. Finally, because saving rates have been exceptionally high in the region, they account for an even more rapidly growing share of global savings. The last two points have been weakened by the recent Asian economic crisis, but nonetheless remain valid. In a mechanical extrapolation exercise similar to ones for the OECD and European Union cited above, Heller and Symansky (1997) concluded that after 2025, demographic trends in the Asian Tigers will strongly reduce the global supply of savings.

All these analyses are partial equilibrium in nature, meaning that many adjustment mechanisms are not taken into account. As a result, they overstate the negative outlook for savings, investment, and macroeconomic performance. While aging in a region may reduce the supply of savings, slower labor-force growth will also reduce the amount of investment needed to achieve any desired level of output per worker (Cutler *et al.*, 1990; Masson and Tryon, 1990). In other words, while aging is putting downward pressure on the saving rate (s, in Section 3.1.2), the accompanying slower growth rate of the labor force is reducing the amount of investment required for capital deepening (rk in Section 3.1.2). Whether current population trends ultimately raise or lower capital per worker and living standards is an empirical question. Changes in interest rates, exchange rates, and capital flows will work to eliminate economic scarcities that arise from an older age distribution and to soften the adjustment process. Changes in wage rates associated with growing scarcity of labor will have impacts on age-specific labor-force participation rates and may affect men and women differently (Fair and Dominguez, 1991). While the strength of the offset is disputed, there is evidence that when individuals recognize the future impacts of population aging on public-sector deficits, they will increase their own saving to compensate for the effect (Barro, 1974; Buiter, 1988). Finally, government policymakers are unlikely to remain passive in the face of deteriorating fiscal accounts (MacKellar and McGreevey, 1999). Pension and health systems can be reformed, labor markets can be made to operate more efficiently, etc.

A number of analyses have tried to take these adjustment and response factors into account. The most comprehensive such study was done by the OECD (Turner *et al.*, 1998), which used a multiregional linked model of the global economy to simulate the effects of slower labor-force growth and population aging out to 2050.[16] Their reference scenario projected a generalized slowdown in the rate of improvement in living standards due to less favorable demographic trends. A range of simulated policy responses (higher retirement age, fiscal consolidation in advance of the baby boomers' retirement, etc.) were estimated to have the potential to soften the adverse impacts of population aging, but only a powerful combination of simultaneous measures could completely eliminate them.

3.5 Conclusion

The main conclusion of a simple neoclassical model of population and economic development is that, while there is no relationship between the rate of population growth (and thus high fertility) and the rate of per capita income growth, there is an inverse relationship between the rate of population growth and the level of per capita income. However, barring market failure there is no reason to believe that high fertility combined with low income is not a utility-maximizing choice. In addition, research has provided little support for the widely held belief that the macroeconomic effects of high fertility and rapid population growth must inevitably mire nations in poverty.

The basic neoclassical model can be enriched by taking into account household- and community-level interactions between high fertility, poverty, low status of women and children, and environmental damage. Such vicious-circle arguments have become a powerful rationale for population policies. They incorporate flexibility, in the form of neoclassical responses to scarcity; equity, by recognizing that poverty, insecurity, and low status of women and children can impede such responses; and ecology, by stressing the fragility of marginal environmental zones. In extending the market-based neoclassical model to include vicious-circle elements, researchers and policymakers recognize that markets must function in an often hostile social and historical context. Attention centers on those populations that are simultaneously most likely to suffer consequences of climate change and least able to adapt to it. And finally, the discussion focuses on what sorts of policies strengthen the ability of institutions of all kinds to cope with environmental stress.

While many case studies have described interactions between high fertility and well-being, systematic, context-independent empirical evidence is not overwhelming. Few studies go beyond correlations, and those that do find causal relationships that are significant but modest. Careful researchers have emphasized the role that institutions play in mediating between demographic pressure and negative impacts. The central message of the research presented here is, however, that high fertility has at least some significant adverse household- and local-level impacts. In addition, individual fertility decisions resulting in rapid population growth carry an external cost (including a GHG emissions cost imposed on future generations) in the form of impaired supply of public, open access natural resources and environmental services.

When population aging and the accompanying slowdown in labor-force growth are added to the picture, a new set of complications arises. Much research suggests that population aging, especially in MDCs, may squeeze the global supply of savings; however, slower labor-force growth will simultaneously reduce investment needs. While aging has the potential to weaken the world economy, it can be countered by appropriate policy responses. However, it is striking that the challenges of global climate change and global population aging will have to be confronted at the same time.

Notes

[1] In a closed economy, or for the world as a whole, net savings (gross national product minus consumption minus depreciation) are by definition equal to net capital formation, or investment. Any *ex ante* imbalance between planned saving (supply of funds) and planned investment (demand for funds) is resolved by a change in interest rates and the level of economic activity. In an open economy, investment is equal to net savings from domestic incomes plus the balance on current transactions with the rest of the world. In the case of a current account surplus, domestic savings are exported to finance investment abroad in return for a claim on output (i.e., net foreign assets rise). In the case of a current account deficit, "foreign savings" flow into the country to supplement domestic resources in financing investment, in return for which foreigners accumulate a claim on future output (i.e., net foreign assets are reduced). In the open-economy case, *ex ante* imbalances between planned investment and saving are resolved by a combination of changes in interest rates, GDP, exchange rates, and net foreign assets. The fact that different world regions are experiencing different demographic dynamics is likely to give rise to large international movements of capital (Turner *et al.*, 1998; Börsch-Supan, 1991; Masson and Tryon, 1990; Cutler *et al.*, 1990). Changes in global economic variables as a result of such flows are, however, likely to be modest (Turner *et al.*, 1998; MacKellar and Reisen, 1998).

[2] The first set of possible effects may be conveniently called "Boserup–Simon effects," after their two most famous proponents (Boserup, 1965; Simon, 1981). While there is general agreement that these are a major force in agriculture, their relevance for manufacturing is less certain (let alone their significance as an economy-wide proposition, as argued by Simon). The second set of effects, by which demographic factors may retard or accelerate technical progress, is close to the heart of endogenous growth theory touched upon later.

[3] It is common in specifying long-run growth models to employ a Cobb–Douglas production function, in which the constant term grows exponentially at some rate – say 2% per year – representing technical progress. Since this is a multiplicative function, the growing constant acts like a rising tide that lifts all ships. Simon specified the constant term to be $1.0 + 0.5\ r$, where r was the growth rate of the population (Simon, 1976).

[4] In simple univariate regressions employing cross-country observations, a one percentage point difference in the population growth rate was found to be positively associated with a two percentage point difference in the rate of growth of the economically active population in agriculture (Gilland, 1986) and a one-third percentage point difference in the rate of expansion of cultivated area (Mink, 1993).

[5] The economic theory of natural resource markets and the economic theory of the environment are identical; we use the two terms interchangeably.

[6] The result was derived for a nonrenewable resource but is easily generalized to cover renewable resources as well. "Stumpage fees" in forestry provide an example of a scarcity rent in the renewable resource domain.

[7] A less restrictive and hence less "eco-friendly" definition is that the value of the entire capital stock (natural, human, and manufactured) should not diminish in value over time. This allows for substitution between the various forms of capital.

[8] The classic study in this area (Lee, 1980) found a strikingly high correlation between the rate of population growth and the rate of increase in the price of grain (which is, in a preindustrial society, effectively equivalent to the price of land) in England between the 13th and 19th centuries.

[9] The pre-analytic vision of ecological economics is so heavily informed by thermodynamics that it might be called "thermodynamic economics" with little loss. The circular-flow, sources–uses accounts that are key to neoclassical economics ignore entropy losses; that is, they assume that the economy is a closed system. When the global economy was small relative to the ecosystem, argue ecological economists, such an assumption could be maintained. Now that the scale of human activity is pressing up against limits, failure to take account of entropy losses is giving rise to spectacularly wasteful natural resource consumption patterns – patterns that presage disaster unless they are scaled back.

[10] For example, Daily *et al.* (1994) estimate that the optimal world population is 1.5–2 billion people; among the factors considered are that the population should be large enough to support urban centers, which provide a critical mass for the arts, yet small enough to allow indigenous cultures breathing space to live in isolation. This example in itself shows how contingent carrying capacity is on cultural norms and values: it is based on a particular way of seeing human beings and their needs and wants, as well as a given view of the ability or inability of human beings to deflect and modify impacts on the environment through institutions.

[11] The private discount rate (as opposed to the social discount rate, discussed in *Box 3.1*) consists of two components: a pure rate of time preference, which is generally held to be the same across individuals, and a risk premium, which reflects mainly risks of expropriation. For those living in security, the risk premium is small; for those living in perpetual insecurity, it is so large as to swamp the pure rate of time preference. The link between poverty and high discount rates is a theme developed by Pearce *et al.* (1990).

[12] The example is meant only to be illustrative. Bradley and Campbell (1998) have argued that wood fuel shortages are not necessarily linked to deforestation.

[13] A similar cautionary note was sounded by Wolf and Ying-Chang (1994, p. 433) in the case of female labor-force participation. In summarizing their finding that fertility in a district of rural Taiwan from 1905–1980 did not vary according to women's agricultural labor-force participation, they write, "It does not suffice for women to be involved in labor if they are involved only as laborers. Their fertility is only affected when women control what they produce." Once again, attention seems to have

been focused on an easily quantifiable variable, in this case, the female labor-force participation rate.

[14] The standard of living; the availability of untapped, potentially cultivable land; the availability of off-farm employment; the potential for labor-intensive, land-saving technological change; the crop structure and its capacity to change; the size of the rural population relative to the urban population; the level of rural fertility and strength of factors determining its persistence; the existing size of landholdings and their distribution; and the existing institutional structure.

[15] The hump-shaped age profile of saving over the lifetime emerges from the so-called life-cycle hypothesis (LCH), according to which individuals smooth their consumption path by reducing consumption when they are young (i.e., saving) in order to increase consumption when they are old. Addition of bequest motives, liquidity constraints (the inability of young persons to borrow against expected future income), precautionary motives (saving against the contingency of ill health), etc., complicates the basic model considerably. While empirical evidence is mixed (cross-sectional data support the LCH more strongly than time-series data, macroeconomic analyses are more supportive than microeconomic studies), the bulk of it supports at least some weakened version of the LCH.

[16] The OECD simulation, as well as other model-based simulations of population aging and the global economy, treat demography in a rather cursory fashion in that they do not contain full cohort-component population projection modules, but rely instead on summary measures such as the aged dependency ratio. On the other hand, simulations at the national level using overlapping generations (OLG) models, which embody detailed demographic accounts, have also generally supported a pessimistic view of population aging (Auerbach *et al.*, 1989; Auerbach and Kotlikoff, 1987). To date, computational difficulties have discouraged researchers from building a linked multiregional OLG model covering the major actors in the world economy.

Part II

Chapter 4

Population and Greenhouse Gas Emissions

As discussed in Chapter 1, if greenhouse gas (GHG) emissions are not constrained, the global average temperature is likely to rise 1–3.5 degrees Celsius (°C) by 2100. About half that range is due to uncertainty in the future unconstrained emissions path (i.e., assuming no policies aimed at reducing emissions are put in place), and the other half is due to uncertainty in the response of the climate system. Future emissions are likely to be an even more important determinant of future climate change, however, since the range of potential emission paths is considerably widened by taking into account policies that could reduce emission rates.

Broadly speaking, demographic change, changes in economic output, and changes in the GHG intensity of the global economy are the forces driving GHG emissions. Each of these is, in turn, influenced by a number of important indirect variables. Regarding the role of demographic change, Chapter 2 demonstrates that a wide range of population paths is possible and that the primary determinant of future population size and structure will be trends in fertility rates.

Taken together, these observations suggest that by slowing population growth, policies that tend to reduce fertility could contribute to reducing emissions and averting climate change. In this chapter we address the questions of how much such policies might reduce GHG emissions and how these reductions would compare with reductions achievable through other means. We discuss the human activities that give rise to GHG emissions and review studies that have used demographic impact identities based on the Impact–Population–Affluence–Technology (I=PAT) equation to apportion responsibility for emission trends among driving forces. We

113

conclude that the results of I=PAT decomposition exercises are difficult to interpret and compare, and therefore provide little guidance for policy.

However, the I=PAT equation can serve as a useful organizing perspective for GHG emission analyses, and it is a logical starting point for generating and comparing alternative emission scenarios. We use our own version of the equation as a simple model to test the sensitivity of GHG emissions to reductions in the rate of population growth. We consider first the direct scale effect of population on GHG emissions. We then modify the model to explore the role of households, as distinct from individuals, in generating emissions and to examine indirect effects such as the possible link between the rate of population growth and the rate of growth of per capita gross domestic product (GDP). We also review a number of influential climate change assessment models (among them the family of models underlying the scenarios produced by the Intergovernmental Panel on Climate Change, or IPCC) and conclude that their treatment of the demographic variable has not gone far beyond the I=PAT equation.

Our central conclusion is that the direct effect of policies to accelerate demographic transition is to reduce GHG emissions significantly in the long term. In the nearer term, through the middle of the 21st century, effects are much smaller. While assumptions such as the demographic units employed (e.g., individuals versus households) may be significant factors in projecting absolute levels of emissions, they do not change the sensitivity of the model to alternative population paths very much and therefore leave the central conclusion of the exercise unchanged. Reasonable parameter estimates suggest that including possible indirect effects also does not change model sensitivity to population enough to reverse the basic direct scale effect of population. Induced effects, by which population growth affects responses by existing institutions and the evolution of new institutional forms, are a different matter, but we include only a limited discussion of them here. Chapters 3 and 5 discuss such impacts in more detail.

The finding that the total (direct plus indirect) effect of slowing population growth is likely to alleviate the problem of GHG emissions does not address the more relevant (and challenging) question of how the costs and benefits of policies related to population compare with the costs and benefits of other policies, such as those to develop alternative technologies. We discuss this question further in Chapter 6.

4.1   Population, Consumption, and Greenhouse Gas Emissions

Human beings consume natural resources through their consumption of goods and services. Pollution, including emissions of carbon dioxide (CO_2) and other GHGs, arises because production of goods and services (and often their consumption, as well) gives rise to residuals. Production of goods and services ultimately depends on consumer demand; even government and military expenditures, for example, ultimately depend on the demand of households for public services and security. Some goods, such as fresh fruits and vegetables, are consumed directly; others are intermediate goods used for the production of goods and services that are subsequently consumed directly (steel used to produce cars, for example, or electricity used to light office buildings); others are investment goods used to build up the capital stock necessary to produce consumption and intermediate goods.

The point is that, at the end of the line, there must be human consumption of goods and services and hence a human choice. If not, no production, and therefore no utilization of natural resources or emission of pollution, will take place. The role of consumption as prime mover, sometimes called the assumption of "consumer sovereignty," is crucial to the neoclassical economic model of the environment, because it permits economists to argue that consumers make intelligent trade-offs between consumption of goods and the quality of the environment (see Chapter 3).

Of course, production is driven not only by the kinds of choices made by consumers, but also by the scale of consumer demand, determined in part by population size. If fertility is also subject to human choice, and if there is an inverse relationship between the population and the quality of the environment as well as between the population and the material standard of living, then an implicit neoclassical population–environment–consumption trade-off is defined. Even if the methodologically individualistic neoclassical model of human choice is rejected in favor of a different model (based more closely on cultural values, say), and despite the fact that social scientists, philosophers, and humanists may differ on how choices are defined and made, the crucial element of choice remains.

Most of population growth's impact on GHG emissions will consist of its impacts on the consumption demands that give rise to demand for energy (particularly fossil fuels and traditional biomass fuels), agricultural products, and goods whose production is associated with deforestation. Population therefore takes its place among the various other factors that also affect consumption demand: the level of income, relative prices, the interest rate, the nature of property rights, etc., not to mention social structure, culture, values, and tastes. To this list must be added the nature of the basket of goods available, which depends not only on consumer demand, but on the state of technology as well. As population is only one of many factors, the situation is complex.

Direct, Indirect, and Induced Effects

Confronted with this complex reality, a normal scientific response is to search for simple models that broadly summarize the impact of population on availability of natural resources and the quality of the environment. A logical first step in this direction is assessing the scale effect of population, that is, the impact of population independent of possible feedback effects on other relevant variables. Keyfitz (1992) calls this the "direct" effect of population growth on the environment and, citing ambiguities and disputes regarding the nature of indirect effects, argues that it should be the focus of attention. Downplaying the direct linear scale effects of population while stressing its less understood indirect nonlinear effects, Keyfitz argues, can result in defining "the population problem" out of existence. This direct effect was the prime concern of Malthus, and it underlies many contemporary concerns (Daily and Ehrlich, 1992; Daly, 1991).

Indeed, the literature on indirect effects has failed to arrive at particularly strong conclusions. An encapsulation of research findings regarding population and economic growth, agriculture, and deforestation is useful here:

- As reviewed in Chapter 3, the relationship between population and the level of per capita income is ambiguous or, more accurately, depends on a broad range of contextual variables. Two opposing views – first, that demographic growth mires populations in poverty and, second, that demographic growth stimulates improvements in technology and organization – can be traced back as far as the 18th century. A large postwar literature on population and development has failed to reach strong conclusions one way or the other.

- Slower population growth is associated with a higher ratio of the elderly to the working-age population (see Chapter 2), which may be associated with lower availability of savings (see Section 3.4). It is plausible that extreme population aging might "crowd out" global investment in climate change mitigation and adaptation. However, the deceleration of labor-force growth accompanying population aging reduces the need for investment in plants and equipment, which would free up resources for environmental investment. Moreover, in LDCs characterized by the most rapid population growth, fertility decline would reduce the youth dependency ratio, which some research indicates might increase the availability of savings.

- Population growth increases the share of production accounted for by agriculture and the proportion of the land base devoted to food production. In this sense, demographic increase has the effect of concentrating output into a heavily polluting sector. The theme of virtuous and vicious adjustments to population pressure against the natural resource base was derived for the most part from research on population and food. Again, postwar research (reviewed in Chapter 5) offers few clear-cut conclusions on the role of population.

- Most tropical deforestation occurs in a limited number of "hot spots" and can be traced ultimately to expansion of agriculture. Largely because of the link to agriculture, it has long been recognized that rapid population growth is cor-related with rapid loss of forest cover. The correlation says nothing about the more underlying causal relationship, which, as we discuss in Chapter 5, is complex.

These rather equivocal results on the indirect effects of population growth on the global environment support the case for starting off with a simple linear model.

In addition to direct and indirect effects, there is a family of impacts that might be called "induced effects": changes in the way social institutions cope with eco-logical stress, including changes in the institutions themselves, as a result of shifts in demographic regime. Whereas social science research on indirect effects is abun-dant and equivocal, research on induced effects is sparse and, with some notable exceptions (e.g., McNicoll, 1990; Tiffen and Mortimore, 1992), speculative. Under some circumstances, it appears, rapid population growth gives rise to social and institutional changes that promote environmentally sound adaptation; under oth-ers, it gives rise to institutional and social gridlock, with ensuing environmental degradation.

4.2   Demographic Impact Identities

The tool of choice in estimating the direct effects of population growth has been the demographic impact identity. These multiplicative identities express total environ-mental impact as the product of population and per capita environmental impact:

Impact = population × impact per capita. (4.1)

"Impact" is usually taken either as utilization of a natural resource or emission of a pollutant.[1] Impact per capita is generally considered to be a function of economic output and impact per unit of output produced:

Impact = population × output per capita × impact per unit output. (4.2)

In economic accounting, the value of output equals total income (i.e., sources and uses must balance), so output per capita is conventionally referred to as "afflu-ence." Impact per unit of output conventionally referred to as "technology," depends on the nature of the production process employed. Putting all this together, the most common demographic impact identity is the I=PAT equation:

$$I(t) = P(t) \times A(t) \times T(t),$$ (4.3)

where I is natural resources utilized or pollution generated ("impact"); P is population; A is per capita output ("affluence"); T is natural resources used or pollution produced per unit output ("technology"); and the argument t indexes the time dimension.

An example of an I=PAT-type identity might be

$$CO_2 \text{ emissions} = \text{population} \times \text{GDP per capita} \times CO_2 \text{ emissions per unit of GDP.} \quad (4.4)$$

The I=PAT identity was developed in the early 1970s during the course of a debate between Barry Commoner, who argued that environmental impacts in the United States were primarily due to postwar changes in production technology, and Paul Ehrlich and John Holdren, who argued that all three factors were important and emphasized in particular the role of population growth. Ehrlich and Holdren (1970) elaborated an early form of the I=PAT argument to rebut assertions by Commoner and others that population growth was unimportant in explaining rising pollution levels in the United States. Both Commoner (Commoner *et al.*, 1971) and Ehrlich and Holdren (1971, 1972) then formalized the I=PAT equation to make quantitative arguments on the relative importance of the factors contributing to pollution.

A variety of forms of the I=PAT identity have been used to analyze a wide range of issues, including automobile pollution (Commoner, 1991), fertilizer use (Harrison, 1992), energy (Pearce, 1991), and air quality (Cramer, 1998), to name just a few. For any impact, the analysis can be made more or less detailed by breaking down variables on the right-hand side of the identity into further multiplicative terms. For example, fertilizer use might be expressed as population multiplied by food production per capita multiplied by fertilizer use per unit of food produced. Fertilizer use per unit of food produced could be further partitioned into cultivated land area per unit of food produced multiplied by fertilizer use per unit of cultivated land.

I=PAT has been useful in these analyses as an organizing perspective, dividing the driving forces of environmental impact into a manageable number of broad categories. In the fertilizer use example above, population, diet, and the characteristics of the food production system are seen as the main forces. The variables also define broad areas for potential policy. The general I=PAT identity indicates that global environmental pressures can be alleviated by implementing policies to slow population growth, to reduce or slow the growth of consumption per capita (taken as a measure of affluence), or to decrease the environmental impact of production. Usually, reducing consumption growth is considered a potential policy only in more developed countries (MDCs), since stifling the growth of consumption per capita in less developed countries (LDCs) is hardly an option. The combined effect of all these policies must therefore be sufficient to allow for improvements in LDC living standards.

The I=PAT identity also illustrates an important consequence of the multiplicative relationship between driving forces: each variable amplifies changes in any other. As a result, a given change in technology may have only a small effect on the environment in a society with a small, low-income population, while the same change would have a much greater effect in a populous, affluent society. Likewise, a given increment in population would have a much greater impact in affluent societies than in low-income countries, assuming levels of technology are similar.

However, the I=PAT formulation has been strongly criticized. It is often pointed out that the identity implicitly assumes that there are only three relevant variables (broadly defined), and that they are related in a simple linear fashion. Many I=PAT analyses are accompanied by caveats about the simplistic representation of a more complex reality.[2] However, authors sometimes go on to draw unqualified policy-related conclusions nonetheless. Thomas (1992), for example, criticizes for being overly simplistic a study by the United Nations Population Fund (UNFPA, 1991), which makes broad policy recommendations based largely on I=PAT analyses. Some researchers have attempted to put the I=PAT framework into a larger perspective. Shaw (1989) proposes that the P, A, and T variables be thought of as proximate (direct) causes of environmental impact that are themselves influenced by a wide range of indirect, but more fundamental, ultimate causes including warfare, poverty, urban bias in development expenditures, distortionary fiscal and pricing policies, poor land-management policies, and protectionism. Similarly, Harrison (1994) describes a systems view in which P, A, and T are "frontline" variables embedded within a network of social, political, and economic influences. These nuances do not invalidate the I=PAT approach for all applications, but serve as reminders that the equation merely skims results off a much deeper, more complex set of relationships.

Decompositions

The I=PAT equation was developed not only as an organizing perspective, but also as a tool to quantify the relative importance of each of the driving variables. The usual methodology is to decompose historical or projected trends in environmental impacts into contributions from trends in each of the driving variables. But while such decomposition exercises have provided grist for the population–environment debate, they have done little to help resolve it. For example, the publication of Commoner's book *The Closing Circle* (1971), which detailed the argument that changes in technology had been by far the most important determinant of pollution in the United States, set off a bitter dispute between Commoner and Ehrlich and Holdren. The exchange, which appeared in the pages of the *Bulletin of the Atomic Scientists* and was excerpted in Marden and Hodgson (1975), was fueled by decomposition exercises that reached contradictory conclusions. Both sides long held to their original positions, continuing to employ I=PAT decomposition exercises in their defense (Commoner, 1991; Ehrlich and Ehrlich, 1990).

In order to avoid problems inherent in comparing final- to initial-year levels using ratios or percentage changes, it is now standard practice to convert the multiplicative I=PAT model to an additive one (Commoner, 1972; Holdren, 1991). The transformation can be made either by taking the logarithm of both sides of the equation or, as is more commonly done, by differentiating and expressing in terms of growth rates so that I=PAT becomes

$$I' = P' + A' + T',\tag{4.5}$$

where prime notation denotes a continuous growth rate. In practice, growth rates are often calculated in discrete, not continuous, terms, in which case the additive expression remains a good approximation as long as the growth rates are not too large.[3] "Normalization" of decompositions allows comparison between scenarios, time periods, regions, or pollutants. The standard normalization for decomposition of growth rates is to divide all the growth rates by the growth rate of impact, that is, for the I=PAT model,

$$1 = P'/I' + A'/I' + T'/I',\tag{4.6}$$

and to frame the discussion of results in terms of the "percentage contribution" of each variable to impact growth.

The growth rate decomposition methodology has been applied by several researchers to data on GHG emissions. Probably the most widely cited study is that of Bongaarts (1992). Working from the US Environmental Protection Agency's "No Response" scenario (Lashof and Tirpak, 1990), Bongaarts concluded that 50% of the growth in global CO_2 emissions from fossil fuels between 1985 and 2025 was due to population growth. Over the entire simulation period (1985–2100), population growth accounted for 35% of growth in CO_2 emissions. Other authors (Holdren, 1991; Harrison, 1992; MacKellar *et al.*, 1995) have used the same method on similar data sets, arriving at a range of conclusions on the contribution of population growth to growth in energy consumption or GHG emissions.

However, growth rate decompositions suffer from a number of inherent weaknesses that make them difficult to interpret and nearly impossible to compare.[4]

The Offset Problem

If one of the variables on the right-hand side of the I=PAT formulation is shrinking over the time period in question, it will offset the contribution of one of the growing variables, and the third variable will be left apparently accounting for a very large proportion of total environmental impact. This problem often arises in decomposing GHG emission trends, since T (in this case, carbon intensity of GDP) is projected to fall in most regions of the world while both P and A are expected to rise. Many authors (Holdren, 1991; Bongaarts, 1992; MacKellar *et al.*, 1995; Raskin, 1995) sidestep the problem by collapsing A and T into a single, more slowly growing "per capita impact" term, but this approach artificially inflates the importance of population and discards important information about the more rapid trends of consumption growth and declining resource use per unit output. For example, in the Bongaarts (1992) decomposition of CO_2 emissions over the period 1985–2100, whereas the contribution of population was +35%, the contribution of A was +191% and that of T was –126%. The mathematics may be correct, but the language is ambiguous. The reader who focuses only on the 35% contribution of population is left with a misleading overstatement of the importance of population, because the remaining 65% is the net of two changes which are much larger in absolute terms while differing in sign.

Heterogeneity Bias

Demographic heterogeneity may bias decompositions in two ways. The first source of bias arises because the population growth rate may be correlated with per capita impact (Lutz, 1993). In the case of global carbon emissions and population growth, because per capita emissions are lowest where population growth is highest, decompositions at the global level overstate the contribution of population growth. Bartiaux and van Ypersele (1993) point out that a number of works that are influential in the public debate, among them Ehrlich and Ehrlich (1990), Gore (1992), and United Nations Population Fund (UNFPA, 1991), make mistakes that arise from aggregating heterogeneous regions. A second source of bias, which might also be considered an offset problem, arises when the signs of the growth rate of a given variable differ across regions. When such regions are aggregated, the variable may appear to be unimportant because changes in the subregions cancel each other out.[5] For example, if T is rising in region A and falling in region B, it may appear at the global level to have undergone no change.

The first source of bias can be minimized by calculating results at the most disaggregated level possible. One problem this raises is that any subsequent aggregation of these regional growth rates into a global growth rate requires a weighting method, and a range of methods, all yielding different results, is available.[6] Canceling, the second source of bias, is more difficult to avoid in aggregating results, but its significance can be measured by examining the regional data directly.

Absolute versus Relative Change

It is the absolute amount of pollutant emissions that damages the environment, not annual growth rates of emissions (Keyfitz, 1992). Framing the discussion in terms of dimensionless growth rates is dangerous if there are ecological discontinuities, nonlinearities, and threshold effects. It is impossible to decompose absolute change in multiplicative identities in the same neat, additive way as growth rates.

What Kind of Growth Rate?

Some authors abandon annual growth rates and instead judge the relative importance of variables according to their effect on impact growth over the entire period. One method, which has been applied to CO_2 emissions (Bartiaux and van Ypersele, 1993) and methane (CH_4) emissions (Heilig, 1994), is to freeze the variable of interest at its initial value and calculate how much the growth in total impact is reduced as a result. Another method, which has been applied to energy demand (Howarth *et al.*, 1991; Ang, 1993) and CO_2 emissions (Moomaw and Tullis, 1998), is to freeze all variables except the one of interest at their initial values and calculate the resulting growth in impact. The two methods produce different results.

Approximation Methods

The standard additive growth rate decomposition expression is exact only for continuous rates of change. For discrete growth rates, it is an approximation obtained by dropping terms of order 2 and higher. The full growth rate decomposition of the I=PAT model, decomposed in average annual percent change terms, is

$$I' = P' + A' + T' + P'A' + A'T' + P'T' + P'A'T'. \tag{4.7}$$

If the growth rates are rapid enough, the interaction terms are too large to neglect. Raskin (1995) has proposed an ad hoc additive decomposition that assigns the cross-terms proportionally to each of the variables based on their total percentage growth. One problem with this method is that it underestimates the contribution of shrinking variables, because the total growth rate for a declining variable is bounded at –100%. A second problem is that, when extended to more than two variables, the Raskin decomposition method violates basic additivity assumptions. For example, it can be shown that the proportion of growth assigned to population in the I=PAT model is not equal to the proportion assigned in the model I=P × [AT], where the last two variables in brackets have been collapsed into a single variable.

Alternative Normalizations

Some of the problems with standard growth rate decompositions, in which growth rates of driving forces are normalized relative to the growth rate of the impact variable, can be ameliorated by using an alternative normalization. But solving one problem often creates another. For example, Harrison (1992) tries to solve the problem of declining variables by first classifying variables as either increasing or decreasing. The growth rate of any particular variable is then normalized to the sum of all growth rates in its sign class. For example, if P, A, and T are growing at +1%, +2%, and –2%, respectively, then their respective contributions to the increase in impact are +33%, +67%, and –100%. This method gives an inaccurate impression of the relative importance of each variable. The increase in A in this case is enough to completely offset the decline in T, yet it appears to be only two-thirds as important. In addition, the contributions of the three variables sum to zero instead of to 100%.[7]

To summarize this discussion, there is a range of decomposition methodologies, each with its own shortcomings, and results depend on the method used. Furthermore, this list of difficulties by no means exhausts the problems associated with decomposition exercises. One of the most important is that all decompositions implicitly measure the importance of a variable relative to a scenario in which it is not growing at all – an assumption that is at best implausible and at worst impossible. For example, demographic momentum ensures some additional population growth. In addition, decompositions are sensitive to two issues we take up in more detail later in the chapter: choice of demographic unit (if the number of households grows faster than population, decompositions done using the I=PAT equation will assign less responsibility to demography than decompositions done using a modified model incorporating households), and possible relationships between right-hand-side variables (if population growth is inversely correlated with per capita income growth, for example, decompositions will assign too much responsibility to P and not enough to A).

Taken together, all the difficulties associated with decompositions make their results of little value in assessing the importance of population policies relative to other policies to reduce GHG emissions. This is not to say, however, that the I=PAT framework should be discarded. While decomposition exercises have not shed much light on the role of population growth in GHG emissions, the application of I=PAT to sensitivity analysis is more promising. By using the I=PAT framework as a simple linear model – a useful initial approach to a complicated problem – the impact of plausible alternative scenarios for the driving forces can be compared. Such sensitivity analyses avoid many of the problems associated with decomposition exercises, and because they simulate the effect of alternative policies, they are more relevant to policy questions. In the next section, we use an I=PAT model to perform sensitivity analyses of the effect of population growth on GHG emission scenarios. We also modify the model to explore the potential significance of indirect, nonlinear effects.

4.3   Population Sensitivity Analyses

In this section, we simulate the impact on GHG emissions of more rapid demographic transition and contrast it with the impact of lower per capita emissions. The International Institute for Applied Systems Analysis (IIASA) rapid demographic transition scenario represents a future path of population that can plausibly be achieved through policies that encourage lower fertility, such as female education, family planning programs, and social and economic development. As discussed in Chapter 2, this scenario also assumes that it is likely that mortality will decline along with fertility. In the rapid demographic transition scenario, the total fertility rate in MDCs remains low or continues to decline, falling to 1.3 or 1.4 by 2030–2035. In LDCs, fertility declines rapidly and falls to slightly or modestly below replacement level by 2030–2035 (see Appendix I).

We begin the exercise with a linear I=PAT model combining IPCC per capita emission estimates with IIASA population projections and calculate the emission reductions achieved by a switch from the central population scenario to the rapid demographic transition scenario. We compare the results with reductions that would be achieved by a switch from the IPCC's central per capita emission scenario to its low scenario. While the scenarios are not directly comparable (for example, moving from the central to the low population scenario might require expenditure of more or less money than moving from the central to the low emission scenario), they do provide a meaningful comparison of the range of variation implied by alternative demographic and technological futures. We then test the robustness of the alternative demographic scenario by modifying the model's unit of demographic account from people to a combination of people and households. We also investigate the result of incorporating into the model an assumed relationship between the growth rates of population and per capita GDP. Finally, we review the rather limited number of population sensitivity analyses that have been performed with climate change assessment models.

4.3.1   A linear model

Our starting point is the set of emission scenarios constructed for the IPCC using the Atmospheric Stabilization Framework (ASF) model (Leggett *et al.*, 1992; see Chapter 1).[8] We chose these scenarios for their long time horizon (to 2100), comprehensiveness [inclusion of emissions of CH_4 and nitrous oxide (N_2O) as well as CO_2], and widespread use. The more recent projections by IIASA and the World Energy Council (Nakićenović *et al.*, 1998) provide an updated outlook on future energy-related emissions, but switching to these more recent estimates would be unlikely to alter our basic conclusions. To link IPCC scenarios with the IIASA

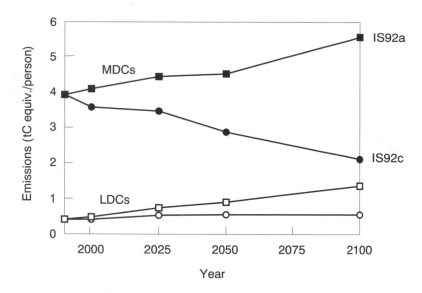

Figure 4.1. Per capita greenhouse gas emissions from commercial energy (IS92a, IS92c), 1990–2100. Source: Pepper *et al.*, 1992.

population scenarios, we combined the IPCC's projections of per capita GHG emissions with the IIASA population scenarios discussed in Chapter 2.[9] This ad hoc strategy is justifiable because the treatment of population in the ASF model is very simple: in most sectors, population simply scales economic activity up or down, with no endogenous feedback of population on per capita income.[10] In noneconomic sectors of the models, such as emissions from agriculture, the relationship between population size and emissions is also direct and linear. Therefore, while total emissions are a function of population (as a scaling factor), per capita emissions are largely independent of population. We focus on emissions from commercial energy, deriving per capita emissions of N_2O, CH_4, and CO_2.[11]

The IPCC constructed six "business as usual" scenarios, which assume that no policies in response to the threat of climate change are put in place. The commonly cited mid-range estimate, IS92a, is taken as our central per capita emission scenario. The lowest estimate, IS92c, is taken as our low per capita emission scenario. *Figure 4.1* shows the two IPCC scenario paths for per capita GHG emissions from commercial energy in LDCs and MDCs. While MDC per capita emissions remain well above LDC rates throughout the 21st century, the ratio of MDC to LDC per capita emissions falls in both scenarios from about 10 in 1990 to less than 4 in 2100.[12] This occurs primarily because in all the IPCC scenarios the ratio of MDC to LDC per capita income is assumed to fall from about 14 in 1990 to about 5 in 2100. If the inequality in income were assumed to be eliminated entirely

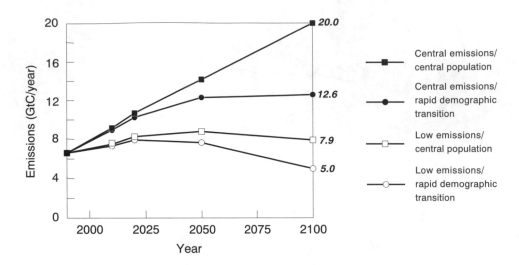

Figure 4.2. World greenhouse gas emissions from commercial energy, 1990–2100.

over the course of this century, a more dramatic shift of economic activity, and thus emissions, toward the LDCs would be expected.

Figures 4.2 to *4.4* show total GHG emissions for the world (*Figure 4.2*), MDCs (*Figure 4.3*), and LDCs (*Figure 4.4*) that result from combining the IPCC per capita emission paths with the IIASA population scenarios using the I=PAT framework. Under the central emission/central population scenario, GHG emissions are expected to rise from 6.6 gigatons carbon (GtC) in 1990 to 14.1 GtC in 2050 and to 20.0 GtC in 2100. The source of emissions will shift decisively to the South. In 1990, MDCs accounted for 75% of GHG emissions from commercial energy; according to the central emissions/central population scenario, MDCs will account for only 47% in 2050 and 39% in 2100.

Sensitivity to Population

Which set of policies – policies to accelerate demographic transition or policies to reduce emissions per capita – would have the greatest effect on emissions? In the near to medium term (up to 2050), the simulations show that policies impinging on per capita emissions have a far greater effect; under the central emission scenario, the difference between emission estimates corresponding to the central and rapid demographic transition scenarios is slim. This result is due mainly to the momentum of population growth, which limits the rate of divergence of alternative population paths, and therefore of emissions as well. In contrast, under the central demographic scenario, the difference between emissions resulting from the central and low per capita emission scenarios is over 5 GtC.

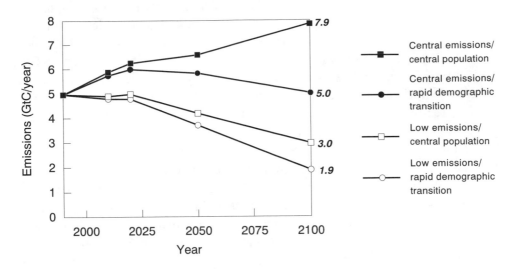

Figure 4.3. MDC greenhouse gas emissions from commercial energy, 1990–2100.

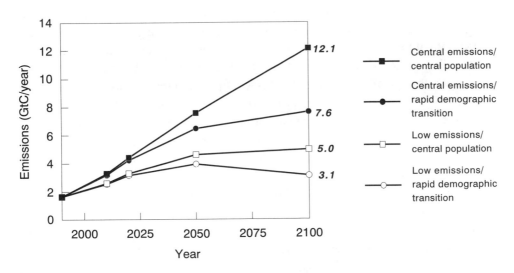

Figure 4.4. LDC greenhouse gas emissions from commercial energy, 1990–2100.

In the long term, however, policies to accelerate demographic transition can make a substantial contribution to slowing growth in GHG emissions. Again under the central emission scenario, in MDCs more rapid demographic transition reduces projected emissions in 2100 by 37%, from 7.9 to 5.0 GtC. In LDCs more rapid demographic transition also reduces emissions by 37%, from 12.1 to 7.6 GtC. Reductions are equal in the two regions in percentage terms because the rapid demographic transition scenario implies equal percentage reductions in population in the two regions in 2100, despite the fact that in absolute terms the reduction in population size is much greater in LDCs than in MDCs. This also implies that the results of the global emission sensitivity analysis do not depend on the ratio of per capita emissions between the regions assumed in the IPCC scenarios; no matter how global emissions are distributed, because population is reduced in equal proportions in both regions, global emissions will always be reduced by the same fraction. This is roughly true at greater levels of regional disaggregation as well. Subregions within the MDC and LDC categories all experience 30–45% reductions in population size in 2100 under the rapid demographic transition scenario, so a more disaggregated analysis would yield similar emission reductions. On the other hand, the greater the MDC-to-LDC ratio of per capita emissions, the greater the proportion of absolute global reductions that will be due to MDC population reductions.

In both MDCs and LDCs, the difference in emissions between the central and low per capita emission scenarios is greater than the difference in emissions between the central and rapid demographic transition scenarios. Nonetheless, it is clear that slower population growth plays a significant role. Moreover, slower population growth in the MDCs accounts for nearly 40% of the global emission reduction (in absolute terms) achieved in 2100, even though virtually all the population growth during the 21st century will take place in the LDCs. High MDC per capita emission rates magnify the impact of MDC population reductions. Taken together, results suggest that, over the long term, slowing population growth can complement or substitute for policies to reduce emissions per capita or per household.[13]

In the rapid demographic transition scenario, the elderly dependency ratio (population over 60 divided by population aged 15–59) is higher for both regions than it is in the central demographic scenario. It can be argued that this might reduce global savings and, by choking off investment, result in higher GHG emissions. A back-of-the-envelope calculation suggests that the difference between the two population scenarios is probably not great enough to support this view.[14] In order to translate population aging into environmental effects via the savings–investment link, it would be necessary to hypothesize either that more rapid aging slows the rate of technological progress, or that it shifts the structure of demand toward more carbon-intensive sectors, or that it leads to fiscal gridlock in the MDCs. Moreover, while evidence is mixed (see Section 3.1.3), some research suggests that the lower LDC youth dependency ratio (population aged 0–14 divided by population aged 15–59) in the rapid demographic transition scenario would substantially raise the supply of savings in these countries (Higgins and Williamson, 1997).

4.3.2   Choice of demographic unit

A basic question relating to the I=PAT model is the choice of the individual versus some other demographic unit of account. In some decomposition exercises – such as the decomposition of trends in the crude birth rate into age-structure and fertility-rate effects, or the decomposition of growth in aggregate economic output into growth of per worker productivity and growth of the labor force – the choice of variables is a straightforward matter of accounting. In the case of I=PAT the choice is less self-evident. Why not, for example, a model such as

Energy use = number of households

$\times$ energy use per household, (4.8)

or

Energy use = population

$\times$ households/per person

$\times$ energy use per household, (4.9)

where the second term on the right-hand side is the inverse of average household size? I=PAT assumes that the environmental impact being modeled arises at the level of the individual, not the household or the community, for example. Yet, to take energy consumption as an example, there are substantial economies of scale at the household level. A significant proportion of residential energy consumption should be assigned to "household overhead"; that is, it is tied to the hearth, not to the number of members of the household.

The possible relevance of alternative demographic denominators points toward a partitioned demographic impact, such as

$$I(t) = D_1(t)A_1(t)T_1(t) + D_2(t)A_2(t)T_2(t) + \dots, \qquad (4.10)$$

where each of the component models is specific to a different sector (residential, industrial, etc.), the Ds are the demographic units most appropriate to that sector (households, workers, etc.), and A and T are appropriately defined to preserve the identity. If the Ds are growing at rates different from that of the aggregate population, either because of shifting age structure or changing behavior, then the partitioned model will likely give rise to results different from those of the I=PAT model. In fact, in a world of rapid population aging and social change, results based on partitioned models may differ significantly from those derived from the simple I=PAT approach.

A Partitioned Model

In Appendix II, we review studies on household-level economies of scale in energy consumption. These lead us to employ a partitioned model approach in which half the emissions are assigned to individuals and half to households:

$$I(t) = 0.5P(t)[GDP(t)/P(t)][GHG(t)/GDP(t)]$$
$$+0.5H(t)[GDP(t)/H(t)][GHG(t)/GDP(t)], \qquad (4.11)$$

where $H(t)$ is the number of households and the last term is the GHG emissions intensity of GDP.

Recent and projected trends in the number of households are discussed in Chapter 2. The essential feature of the projections is that population aging during the coming century will reinforce a historical trend toward smaller average household size, since there will be a smaller proportion of younger individuals (who tend to live with their parents) and a larger proportion of elderly persons living alone. Thus the number of households in both MDCs and LDCs is expected to grow faster than population itself. This result is based on the assumption of constant age-specific headship rates; that is, the fraction of each age group that can be considered to be heading households does not change with time. If, instead, one assumes a continuation of historical trends toward living alone in given age groups (following the concept of continued individuation discussed in Chapter 2), the number of households increases even more rapidly than is assumed here.

Given the IPCC paths for per capita emissions, how does one make assumptions about the path of per household emissions? We have assumed that per household emissions grow at the same rate as per capita emissions; that is, the difference between total emissions under the per capita and per household models arises purely from differences in the rate of growth of the two demographic aggregates. A less ad hoc approach would improve projection results; whether it would change the sensitivity of total emissions to alternative demographic scenarios is doubtful.[15]

Sensitivity to Population

Figure 4.5 compares the results of the partitioned model with those of the standard I=PAT model under the central emission scenario. The partitioned model projects substantially higher GHG emissions than does the standard model. Global emissions in the partitioned model under the central population scenario rise to 16 GtC

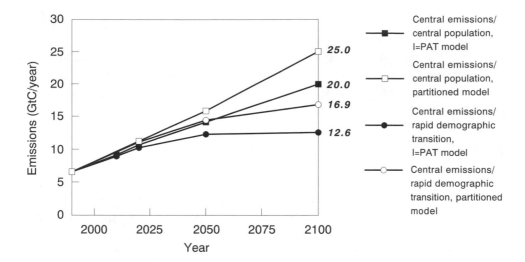

Figure 4.5. World greenhouse gas emissions from commercial energy, 1990–2100, I=PAT versus a partitioned model.

in 2050 and 25 GtC in 2100, 25% higher than in the standard model. This difference is driven by the growth rate of households, which is higher than the growth rate of population in both MDCs and LDCs.

However, the simulations show that, in the near term, the difference between emission estimates corresponding to the central and rapid demographic transition scenarios is small regardless of whether projections are made with the standard or partitioned model. Actually, in the very near term, here taken to be 2020, policies to accelerate demographic transition have no impact at all according to the partitioned model. In part, this is because of the momentum of population growth. Another part of the story, however, is that, whereas the lower fertility in the rapid demographic transition scenario has no effect (relative to the central scenario) on the number of households for at least 20 years (since children are born into existing households), the lower mortality in the scenario immediately starts to increase the number of households consisting of elderly persons. Looked at from the more abstract perspective of changes in the population age structure, lower mortality induces immediate population aging from the top, thus increasing the weighting of high age-specific household headship rates at older ages.

However, the long-term sensitivity to population displayed by the standard model is nearly as strong in the partitioned model. In the MDCs, more rapid demographic transition reduces projected emissions in 2100 by 34%, from 8.2 to 5.4 GtC. In LDCs, more rapid demographic transition reduces emissions by 33%, from 16.9 to 11.4 GtC. The resulting reduction in global emissions by one-third is little different from the 37% reduction projected by the standard model. In other words, considering age-structure effects that increase the number of households does not alter the conclusion that slowing population growth is likely to reduce GHG emissions significantly in the long run. At the same time, however, the marked difference in absolute emission levels between the I=PAT model and the partitioned model points to the importance of accounting for age-structure effects in emission projections.

4.3.3   Relationships between variables

Criticisms of linear models generally assert that the problem has been oversimplified (e.g., Shaw, 1993). If the implied *ceteris paribus* assumptions were replaced with more realistic relationships between variables, the reasoning goes, the results would change.

Among the important *ceteris paribus* assumptions built into the I=PAT model are the following:

- P *is not a function of* A. Populations in rich countries grow more slowly than populations in poor countries. Mortality rates are lower but the gap between fertility rates is even wider, resulting in a much slower overall rate of increase. The relationship between fertility, mortality, and per capita income is far from well understood; indeed, the relationships may be so dependent on conditioning variables as to be nearly incomprehensible (see Chapter 2). But the least likely theory is that there is no relationship at all.
- A *is not a function of* P. In view of the fact that research has failed to uncover a strong relationship between the rate of population growth and the rate of economic growth, this may not be an unreasonable assumption in the relatively near term. Over the long term, however, this assumption denies the possibility of the learning by doing and increasing returns to scale that are at the heart of endogenous economic growth theory (see Chapter 3).
- T *is not a function of* P. Economic structure and technology are, in large degree, responses to population pressure. For example, if population grows rapidly and the supply and rate of renewal of natural resources are fixed by geophysical factors or biogeochemical cycles, then the rising scarcity of natural resources should lead to the adoption of resource-saving or less-polluting production methods; it should also cause consumers to substitute toward less resource-intensive goods and services (see Chapter 3).
- A *is not a function of* T. It would seem likely that the more efficiently materials can be transformed into an economic product, and the less noxious the residuals generated during the process, the higher the level of income. Indeed,

the core result of the neoclassical economic growth model is that, in long-run equilibrium, the rate of per capita economic growth will be given by the rate of technological progress, where the latter is defined as the rate of increase in economic output, the level of all inputs remaining the same (see Chapter 3).

- T *is not a function of* A. Researchers have observed very strong associations between A and T (Chenery and Syrquin, 1975; Maddison, 1989; Pandit and Cassetti, 1989). Partly as a result, environmental impact per unit GDP either declines monotonically with the level of development or follows an inverted-U-shaped path, that is, it first rises, then falls (World Bank, 1992). This path reflects not only changes in economic structure, but also the fact that rising income stimulates demand for environmental quality. The combination of changing economic structure and rising demand for environmental quality defines an "environmental transition" (Ruttan, 1971; Antle and Heidebrink, 1995) in which economic growth is associated with environmental deterioration when a country is poor but with environmental improvement after a critical point in national economic development has been reached (see *Box 4.1*).

However, while lists of possible interactions are easy to generate, formalizing and validating them within a quantitative model is another matter (Schneider, 1997). A handful of studies have attempted to test the linear I=PAT model for GHG emissions against cross-sectional data. Dietz and Rosa (1997), in a multiple regression analysis on 1989 data found that the impact of population size on emissions was roughly linear, if anything becoming disproportionately large for the most populous nations. Affluence also had a roughly proportional effect on emissions up to a transition point around US$10,000 per capita, at which its influence stabilized or even declined. DeCanio (1992) also estimated an I=PAT-type of model based on cross-sectional data from the late 1980s to examine the sensitivity of emissions to alternative scenarios for per capita income. Regression analysis indicated an inverted-U relationship between per capita emissions and per capita income with a turning point at around US$17,000 per capita. As a result, increasing affluence had a less-than-proportional impact on emissions in scenarios for the middle of this century; for example, an increase in the global average growth rate of per capita income from 1% to 1.5% produced income levels that were 46% higher in 2050, while emissions increased only 29%. To the extent that trends in GHG emissions reflect trends in energy use, the inverted-U relationship between per capita emissions and per capita income is mirroring a similarly shaped relationship between per-unit-GDP emissions and GDP per capita. Energy demand is elastic with respect to GDP when economies are rapidly shifting away from agriculture and toward industry, but inelastic with respect to GDP when economies are moving away from industry toward services; in addition, within some sectors, LDCs are substituting

---

Box 4.1. The environmental transition and its critics.

The empirical literature on the environmental transition is growing rapidly, as is criticism of the model. The principal background paper for this aspect of the 1992 *World Development Report* (World Bank, 1992) was by Shafik and Bandyopadhyay (1992), who examined access to safe water and sanitation, urban atmospheric concentrations of particulate matter and sulfur dioxide (SO_2), municipal waste per capita, and CO_2 emissions per capita. In international cross section, the first two declined monotonically with level of income, the second two followed an inverted U-shaped curve peaking at a few hundred dollars in the first case and at just over US$1,000 in the second case. Antle and Heidebrink (1995) found an inverted U-shaped curve relating to deforestation (as did Panayotou, 1993; Cropper and Griffiths, 1994) and availability of national parks; the peak occurred at US$1,200–2,000. Grossman and Krueger (1995) found inverted U-shaped curves for a wide range of water and air pollutants and estimated turning points ranging between US$2,700 (dissolved oxygen content of rivers; this a trough rather than a peak) and US$11,600 (cadmium). According to their results, by the time per capita income reaches US$8,000, levels of virtually all pollutants fall with further economic growth. All these prices are in 1985 Summers–Heston purchasing power parity terms; for reference, GDP per capita in the early 1990s was roughly US$1,000 in India, US$5,000 in Mexico, and US$10,000 in Spain. Most developing countries are far from the income range over which environmental improvement can be expected.

Arrow *et al.* (1995) have expressed strong concern lest the environmental transition model be invoked as a sort of panacea. It is relevant, they argue, only in the absence of ecological feedback and when local social and economic institutions can be counted on to respond to environmental damage. Antle and Heidebrink (1995) are similarly cautious, as are contributors to comprehensive reviews of the literature (Rothman and de Bruyn, 1998; Barbier, 1997; Stern, 1998).

---

energy-intensive production methods for labor-intensive ones.[16] Also contributing to the pattern is the move toward cleaner-burning fuels that occurs as income rises.

To test the importance of indirect effects, we modified the I=PAT model to incorporate potential relationships between the rates of growth of the variables on the right-hand side of the equation. Because we are concerned here primarily with the effect of alternative demographic futures, we limited our analysis to the potential relationships between P' and A'.[17] Moreover, we limited ourselves to first-order

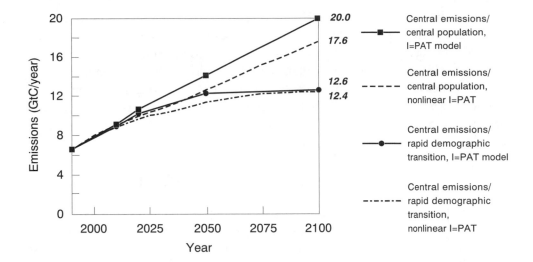

Figure 4.6. World greenhouse gas emissions from commercial energy, 1990–2100, standard versus nonlinear I=PAT.

effects, ignoring, for example, possibilities such as a change in T caused by the change in A arising from a change in P.

Sensitivity to Population

Figure 4.6 compares the results of the standard I=PAT model with the results of a model that assumes that each percentage point deceleration in the rate of population growth increases the rate of per capita economic growth by 0.25 percentage points.[18] In all cases the baseline is the central emission scenario. However, in the nonlinear model, per capita income is altered from the path assumed by the central emission scenario as a result of the assumed relationship between P and A, resulting in slower-growing total emissions whenever population is increasing and faster-growing emissions whenever population is declining.

Both models project similar levels of GHG emissions: in 2100, the central population scenarios produce emission levels that differ by just over 10% (20.0 GtC in the standard model versus 17.6 GtC in the nonlinear model), while the rapid demographic transition scenarios produce emission levels that differ only marginally (12.6 GtC versus 12.4 GtC). The assumed relationship between P and A has almost no effect on long-term emissions under the rapid demographic transition scenario because population actually declines in that scenario over the second half of the century, reversing any divergence in emissions produced by the assumed relationship between P and A while population was growing over the first half of the century.

The projected reduction in GHG emissions in 2100 resulting from slowed population growth remains at a substantial 30% in the nonlinear model (12.4 GtC under the rapid demographic transition scenario versus 17.6 GtC under the central population scenario), not much less than the 37% projected by the standard I=PAT model. Unless the relationship between P and A is far stronger than current research suggests, it seems unlikely that incorporating this particular indirect effect makes a great difference in the sensitivity of emissions to alternative population paths.

4.3.4   Population in climate change assessment models

As one of the driving forces of GHG emissions, population plays an important role in climate change assessment models. In Appendix III we summarize how population is dealt with in nine major climate change assessment models. In general, these models can be divided into two broad classes: projection or scenario models and optimization or endogenous-policy models. In the first group of models, population is treated mostly as a scale factor, the only refinement being the inclusion (via the production function) of some weak ad hoc links between the rate of population growth and the rate of per capita GDP growth – in other words, these models do not progress far beyond the I=PAT models we have used in this chapter.

The optimization models also treat population as a scale factor, on the production and demand sides in Global 2100 (Manne and Richels, 1992; Manne *et al.*, 1995) and GREEN (OECD, 1992), and on the cost and benefit sides in the cost–benefit models of Cline (1992b), Fankhauser (1994), and Nordhaus (1994). By solving for optimal policies, however, this class of model effectively endogenizes induced effects (albeit in a highly simplified form). Consequently, the impact on model results of alternative rates of population growth is potentially more complicated. For example, by increasing the number of people in future generations, higher population growth rates would be expected to raise the optimal degree of GHG emission control in the present; in fact, raising the rate of population growth is conceptually similar to decreasing the discount rate. Further complications may occur to the extent that changing the rate of population growth may affect relative prices, with subsequent effects on production, demand, and the pattern of investment.

Climate change assessment models have not often been employed to simulate the effect of alternative population growth paths (Gaffin, 1998). This section looks at the uncertainty and Monte Carlo analyses of GHG emission models that have explicitly examined population inputs.[19]

Projection results in climate change assessment models vary widely, but this variation is accounted for mostly by differences in exogenous assumptions for

Table 4.1. Ranges of key input variables in sensitivity analyses.

	High/low scenario as % of central scenario, 2100	
	Nordhaus and Yohe (1983)	Edmonds *et al.* (1986b)
Population	+76/–44	+28/–22
Productivity[a]	+202/–32	+2,260/–99
Price of nonfossil fuels	+293/–78	–
Exogenous energy efficiency	–	+1,160/–72

[a]Results from Edmonds *et al.* (1986b) apply to LDCs.

variables such as population, productivity, and exogenous changes in energy intensity. The range of emissions projected by the models for 2100 is reduced from a factor of 40 to a factor of 2 if these and other exogenous inputs are harmonized (Alcamo *et al.*, 1994). A natural conclusion is that uncertainty analyses with respect to such exogenous inputs are at least as important as further refinement of the models themselves.

Early analyses conducted by Nordhaus and Yohe (1983) and Edmonds *et al.* (1986b) found population to be a relatively unimportant source of uncertainty. Nordhaus and Yohe conducted a Monte Carlo analysis on a simple economic–atmospheric model of CO_2 and found that population uncertainty ranked near the bottom of the list of 10 key parameters in explaining uncertainty in atmospheric CO_2 concentrations in 2100. Labor productivity and ease of substitution between fossil and nonfossil fuels were found to be most important. Edmonds *et al.* (1986a) conducted a similar Monte Carlo analysis on the Edmonds–Reilly–Barnes (ERB) model and found that population uncertainty did not even enter the list of the top 10 variables that explain uncertainty over future levels of carbon emissions. Labor productivity was again found to be overwhelmingly important.

As an examination of alternative assumptions makes clear, however, the reason that population ranks low as an uncertainty factor is not because these models are insensitive to population, but simply because demographic momentum makes population a more certain variable than others that appear in the models. *Table 4.1* shows that while productivity and other variables are assumed to have a wide spread of possible values, differences in population size are much more restricted. Indeed, if the input assumptions in *Table 4.1* accurately represent the relative uncertainties, the underlying models would have to be extraordinarily sensitive to population for it to be an important source of uncertainty.

Thus one should not conclude from these studies that the models are insensitive to population. As Edmonds *et al.* (1986b) note, the ERB model is actually relatively sensitive to marginal population perturbations; considering perturbations of all parameters of 1%, population was found to be among the top five most sensitive variables in the model. Similarly, the model used by Nordhaus and Yohe

(1983) should exhibit sensitivity to population that is similar to its sensitivity to labor productivity.

If models are indeed sensitive to population, then where population is compared with variables having similar levels of uncertainty, population will be found to be an important overall source of uncertainty. Such is the case in a sensitivity analysis carried out by Nordhaus on the Dynamic Integrated model of Climate and the Economy, or DICE (Nordhaus, 1994). In contrast to his 1983 study, Nordhaus finds the aggregated output of DICE to be *more* sensitive to population growth assumptions than any other variable, including exogenous productivity growth and the rate of time preference. A high value of future population growth (27 billion by 2100) approximately doubles the rate of optimal GHG abatement in the model and nearly triples the optimal carbon tax (Nordhaus, 1994). Kelly and Kolstad (1996) have shown similar results, outlining in particular the fact that low optimal emission reduction rates in DICE are critically dependent on the assumption of population stabilization.

Gaffin and O'Neill (1997) performed additional sensitivity analyses with the DICE model by investigating optimal CO_2 emission levels assuming population along the UN high, medium, and low paths. They found that optimal emissions in 2150 differed by over a factor of six across these three scenarios, essentially scaling up and down with population. This result argues that, at least within the DICE model, the induced effects described in the previous paragraph do not alter the basic direct scale effect of population on emissions. The authors also examined the contribution that slowing population growth could make to stabilizing atmospheric concentrations of CO_2 as called for by the Framework Convention on Climate Change and found that the low population path made stabilizing concentrations significantly easier.

O'Neill (1996) found the IPCC scenarios of GHG emissions from energy to be quite sensitive to population as well. He fit the expanded I=PAT identity

$$\text{Emissions} = \text{population} \times \text{GDP/person} \times \text{energy/GDP} \times \text{carbon/energy} \qquad (4.12)$$

at the regional level for each set of scenarios. The percentage difference in global emissions due to varying a target variable from its high to low value was calculated for all high/low scenario combinations for the other three variables. The average of these percentage differences was taken as the index of uncertainty due to the target variable. Population uncertainty was found to be significantly more important than uncertainty in either energy intensity or carbon intensity of GNP, although less important than uncertainty over future per capita income. This ordering is consistent through time, and holds in the LDCs and the MDCs alike. In the US Environmental Protection Agency scenarios (Lashof and Tirpak, 1990), which incorporate specific policy efforts to stabilize emissions, the population assumption is less important than the other variables in determining short-term emissions, but by 2100 is again second in importance to per capita income growth in determining emissions. In a similar exercise, Yang and Schneider (1998) identified energy intensity improvements and population growth in LDCs, and GDP growth and carbon intensity improvements in MDCs, as major contributors to emissions in the IPCC scenarios.

4.4 Conclusion

Our central conclusion is that policies to encourage more rapid demographic transition are likely to significantly reduce GHG emissions in the long run. It appears unlikely that age-structure effects (such as the impact of population age structure on the number of households) or indirect effects (such as the possible relationship between the rate of population growth and the rate of per capita income growth) are strong enough to reverse this direct effect. While we arrive at this conclusion based on a very simple model, a number of ad hoc calculations suggest that the result is probably robust; it is also supported by a review of relevant results from climate change assessment models.

Little research exists on the potential role of induced policy or, to put it better, institutional responses. Climate change assessment models incorporate few induced responses that would greatly change the direct scale effect of population on emissions. It is possible that slower population growth would encourage less aggressive GHG abatement policies by reducing the present value of net benefits. However, under low population growth assumptions, unrestrained emission rates are likely to be lower as well.

Of course, many other policies besides slowing population growth would also lower GHG emissions (and could do so more directly, with fewer broad consequences). Whether one would wish to select policies that affect population from a menu of choices is a question we are not yet prepared to answer. We have no common metric for comparing the costs and benefits of policies to slow population growth with those of policies to reduce GHG emissions per capita or per household. We return to this problem in Chapter 6.

Notes

[1] Environmental impact is assumed to occur if a natural resource is being used faster than it is renewed or if a pollutant is being emitted faster than the environment's ability to assimilate it.

[2] Ehrlich and Holdren (1971), for example, used the equation $I=P\times F$, where F was an unspecified function that measured per capita impact on the environment. F was conceived of as a complex function of the level of per capita consumption, the composition of the consumption basket, production technology, and the level of population itself.

[3] For longer time periods, or for fast-growing variables, total growth rates over a period can be too large for the additive relation to remain a good approximation. In such cases, I=PAT can be transformed by expressing each variable as a ratio (R) of its final to initial value over the time period used and taking logarithms. Denoting these ratios R_I, R_P, R_A, and R_T, then

$$\ln(R_I) = \ln(R_P) + \ln(R_A) + \ln(R_T),$$

which yields decomposition results equivalent to those obtained assuming that each variable grew at a constant, continuous rate over the time period.

[4] See Wexler (1996a), Amalric (1995), and Ang (1995) for additional analysis of decomposition methodologies.

[5] This problem is discussed in detail by Ang (1993) in his decomposition of industrial energy demand in Singapore.

[6] The Divisia method, in which the weights applied to P, A, and T are the regional emission shares, is a reasonable choice and ensures that the resulting global growth rates of P, A, and T sum to the actual global GHG emissions growth rate. Every weighting method has its disadvantages, however, and the main disadvantage here is that the emissions-weighted growth rates of P, A, and T at the global level are not equal to the true aggregate growth rates of population, GDP, and emissions intensity. See Wexler (1996a) for application to GHG emissions.

[7] A more mathematically appropriate normalization is to divide each growth rate by the sum of the absolute values of all the growth rates. This alternative normalization method yields a stable set of indices that preserves the relative magnitudes of the growth rates (Wexler, 1996a) and whose absolute values always sum to 100. Thus, the relative importance of each variable can be compared across scenarios, regions, and pollutants. The problem is that in achieving mathematical comparability, the commonsense interpretation of the standard normalization is lost. What meaning is to be attached to the sum of absolute values of growth rates?

[8] Details of the ASF model and other prominent climate change assessment models are given in Wexler (1996b).

[9] We modified the definition of MDCs and LDCs in the population projections of Lutz (1996) by moving Central Asia from the LDCs to the MDCs to agree with the regional definitions used by the IPCC. Sensitivity results (i.e., comparisons of baseline and alternative scenarios) will be virtually unaffected by this modification.

[10] In the energy sector there may be some short-term feedback effect of population on per capita emissions because regional GDPs in the energy models are determined by population above age 15 (Alcamo *et al.*, 1994).

[11] Emissions are given in units of gigatons carbon equivalent (GtC). Carbon equivalent emissions are determined using the 100-year global warming potential (GWP) of each gas taken from Schimel *et al.* (1996); that is, $E_{x,GtC} = GWP_x \times E_X$, where E_X is the rate of emissions of gas X.

[12] Accounting for sources of GHG emissions in addition to commercial energy, such as agriculture and deforestation, the ratio of MDC to LDC per capita emissions in the IS92a scenario remains stable at roughly 4 to 1 over the course of this century.

[13] Two other studies (Kolsrud and Torrey, 1992; Birdsall, 1994a) have concluded the opposite – that GHG emissions are relatively insensitive to population growth. However, their conclusions resulted from the use of relatively short time frames (through 2050) and narrowly divergent population scenarios that do not reflect the full range of potential population paths. Moreover, Kolsrud and Torrey (1992) considered slowing only LDC population growth.

[14] As a first approximation, say that the carbon intensity of output is inversely related to the capital-to-output ratio. This could be because new capital is less carbon-intensive than old and a higher capital-to-output ratio corresponds to a newer capital stock, or because a world abundantly endowed with capital is likely to have resources to spare for carbon mitigation. In the Cobb-Douglas production function $Y = AK^\beta L^{1-\beta}$ the rate of change in the capital-to-output ratio k is given by

$$dk/k = (1 - \beta)(s/k - r),$$

where s is the saving rate ($dK = sY$) and r is the rate of growth of the labor force. An increase in the share of the elderly tends to reduce s, but at the same time is associated with slower labor-force growth r. In the IIASA central population scenario, world labor-force growth between 1995 and 2100 is 0.52% per year; in the rapid demographic transition scenario, it is –0.09% per year. The average elderly dependency ratio (population over 60 divided by population aged 15–59) over the period is 0.328 in the central scenario and 0.496 in the alternative scenario. Say that β is 0.33 and the baseline world net saving rate is 10%. Then, if k started at 3.0 in 1995, it would grow at a steadily decelerating rate to 8.2 in 2100. Now say that, in the rapid demographic transition scenario, the higher elderly dependency ratio has the impact of reducing the average saving rate over the period from 10% to 5%. This would be the result if each percentage point increase in elderly dependency ratio led to a 0.3 percentage point decline in the total (not private) saving rate. Based on empirical evidence regarding private saving rates alone (Turner *et al.*, 1998), this would represent a strong response. With β still equal to 0.33, $s = 0.05$, and $r = -0.0009$, k would grow from 3.0 to a terminal value of 7.0. The difference between 8.2 and 7.0 is not extreme enough to support the view that the more rapid population aging in the alternative demographic scenario would adversely affect emissions per unit of GDP. Slower labor-force growth would tend, *ceteris paribus*, to raise output and GHG emissions per worker but, as we calculate in Section 4.3.3, the impact of emissions, assuming a link between the growth rates of population and GDP per capita, is not very great.

[15] All else being equal, the effect of a decline in average household size is to increase per capita emissions as economies of scale at the household level are diminished. Thus, an emission scenario driven by households of declining average size should reflect faster growth in per capita emissions relative to a baseline scenario that does not account for this demographic effect. Setting the per household emissions growth rate equal to the per capita emissions growth rate, as we do here, does indeed imply faster

per capita emissions growth when households are taken into account. However, it implicitly assumes that the increase in the per capita emissions growth rate is of the same magnitude as the rate of decline in average household size. While this assumption is arbitrary and a more detailed analysis would improve results, the literature cited in Appendix II indicates that it is plausible in quantitative terms.

[16] The elasticity of energy demand with respect to GDP is percent change in energy consumption per percent change in GDP; when this is in excess of 1, demand is said to be "elastic" and when it is between zero and 1, demand is said to be "inelastic." Since GDP equals income, the energy–GDP elasticity can be interpreted as an income elasticity of demand. Data for 1965–1990 from the 1992 *World Development Report* (average annual rate of per capita energy consumption from Table 5, p. 226, divided by average annual rate of per capita GNP growth from Table 1, p. 218) result in estimated elasticities of 1.41, 1.18, and 0.62 for low-, middle-, and high-income countries, respectively (former Soviet Union not included). In an equally ad hoc calculation, Gilland (1988) used elasticities ranging between 0.3 (United States, Canada, and Australasia) and 1.2 (developing countries, not including China and formerly planned economies). Based on an exhaustive survey of energy demand studies in LDCs, Dahl (1994) found that the average long-run income elasticity of demand was 1.27.

[17] Sheffield (1998) examines the implications of an assumed link between the population growth rate and per capita energy use. He derives a relationship between the two variables based on cross-sectional data; when driven by a baseline energy use scenario, the model produces a population growth scenario similar to World Bank projections.

[18] The elasticity of –0.25 is derived from the production function of the DICE climate–economy model (Nordhaus, 1994).

[19] Several important studies (e.g., Grübler, 1994) are based on sensitivity to GDP and fail to disaggregate GDP into GDP per capita and population.

Chapter 5

Population and Adaptation: Agriculture, Health, and Environmental Security

As pointed out in Chapter 1, historical greenhouse gas (GHG) emissions have committed the world to some climate change even if aggressive mitigation policies are put into place immediately. Societies therefore have an incentive to plan for adaptation to climate change impacts no matter how much emissions are reduced. In this chapter, we ask whether slower population growth would enhance the ability of institutions, especially in developing countries, to adapt to the likely impacts of climate change. In 20, 50, and certainly in 100 years, some countries that are currently poor will no longer be so. However, others will, and large subpopulations may remain poor even in countries that attain average living standards much higher than they enjoy today. In considering the impacts of climate change, it is important to impose a hypothetical unfavorable climate not on the world of today, but on the world as it may be decades in the future. Nonetheless, both the geographical distribution of expected climate change impacts and a concern for equity suggest a focus on low-income settings.

We focus on three aspects of human welfare that may be threatened by climate change: agriculture, health, and environmental security. In Chapter 3, we developed a model of population–economy–environment interactions, emphasizing the theme that while stresses to populations in low-income settings can set off a destructive series of effects, policy intervention can encourage virtuous responses as well. Vicious-circle reasoning is also key to our understanding of how population growth affects the ability of societies to respond to the threats posed by climate change in these three areas.

5.1   Food and Agriculture

In the 1970s, the challenge of world population growth was sometimes framed in terms of the threat of mass starvation. In fact, the world food situation is a much more complicated, multidimensional problem. Strictly speaking, starvation is quite rare; however, food scarcity can lead to death from infectious diseases that the body is too weak to resist. Famine mortality in South Asia, for example, consists not of people actually starving to death, but of a massive expansion of normal seasonal mortality patterns, especially mortality due to malaria (Dyson, 1991). Even if it does not result directly in mortality, low caloric intake and/or a poor diet in childhood can have severe long-term consequences for virtually every vital function, and even brief episodes of nutritional stress leave their mark. Poorly nourished people are less productive than well-nourished ones, and adequate nutrition is a vital complementary input to other investments in human capital such as education and primary health care.

Climate change is likely to affect nearly all natural systems and biogeochemical cycles that affect agriculture; as a result, agriculture is the sector placed most directly at risk by climate change. One of the most important direct effects of population growth is to raise the demand for food and fiber; at the same time, one of the most important indirect effects is to influence, through agricultural practices and technology, the way that people provide themselves with that food and fiber. In this section, we summarize the world agricultural situation and review research on the impact of climate change and the role of population in agricultural systems. The last of these should provide us with some indication of how population growth might impede or improve agricultural adaptation to climate change.

5.1.1   Current situation

We have begun to view rising agricultural productivity as a normal state of affairs.[1] Yet, until World War II, most gains in agricultural output at the global level came from expanding the area under cultivation (Wolman, 1993). In the postwar period in contrast, gains have come primarily from rising yields resulting from the use of new seed varieties; expansion of irrigation; multi-cropping; application of fertilizers, herbicides, and pesticides; substitution of machines for labor, etc. The most obvious signs of stress, such as declining per capita output and rising prices, have failed to appear in the world food system.[2] Indeed, many areas are experiencing dramatic improvements in diet.

Yet some authors, such as Lester Brown, have long argued that the world food system is suffering from diminishing marginal yield enhancement and increasing exposure to systemic risk exacerbated by world dependence on the North American grain harvest.[3] Some agricultural economists worry about diminishing returns to the traditional sources of improved yield (Ruttan, 1991, p. 18). Ehrlich *et al.* (1993) have assembled a formidable list of obstacles that must be overcome: loss of farmland to urbanization and nonfarm use, scarcity of fresh water for irrigation, soil erosion, diminishing fertilizer response, problems associated with pests, declining genetic diversity of crops, yield losses to air pollution and increased exposure to ultraviolet-B radiation, impacts of climate change, plus a host of ill-defined problems concerning scarcity of ecosystem services and biological limits to yield increases. Others have expressed concern that, while increasing human populations can be fed, future output gains will come at heavy environmental cost if more ecologically benign technologies are not developed and implemented (Crosson and Anderson, 1994; Engelman and LeRoy, 1995).[4] For example, many yield-enhancing strategies are energy intensive (Chapman and Barker, 1991) and are subject to constraints determined by biogeochemical cycles, such as the nitrogen cycle in the case of fertilizer (Smil, 1991; Kawashima *et al.*, 1997); others have harmful environmental impacts, as in the case of pesticides, herbicides, fertilizers, and some irrigation projects.

A recent study by the International Food Policy Research Institute concluded that, while global food sustainability is an achievable goal, it cannot be realized under business-as-usual conditions (Pinstrup-Anderson and Pandya-Lorch, 1998).

Debate over the current food situation does not end with agricultural potential and environmental impacts, however, because food availability and food security are not necessarily one and the same. Food security requires the elimination of famine, seasonal or chronic malnutrition, micronutrient deficiencies, and nutrient-depleting illnesses. While societies became less vulnerable to famine during the 20th century, the course of undernutrition has been mixed according to region, and trends in micronutrient availability and nutrient-depleting illnesses are ambiguous (Chen and Kates, 1994).

In the entitlements approach pioneered by Sen (1981), food security does not depend on food supply and demand forces per se, but on the distribution of entitlements to food. Those who are integrated into the market economy derive their entitlements mainly from disposable income, while those who are not rely heavily on agricultural land tenure.[5] The rural poor may therefore be affected not only by an increase in the price of food, but also by a further erosion of the already tenuous food entitlements they derive from existing institutional arrangements.

Damage to the agricultural resource base affects food security in three ways. First, it lowers subsistence production directly. Second, by increasing production costs and lowering yields, it decreases farm income. Third, it raises the variance of production, that is, the susceptibility to adverse conditions (such as drought) or disruption (such as floods). Production variance may, in turn, disrupt land tenure and therefore entitlements to food; for example, farmers may be forced to sell land to buy food or, in the worst case, may be permanently displaced. If the declining quality of the resource base results in violent conflict, an entire range of entitlements, including access to transportation and communication networks, will be disrupted.

Achieving food security in the future will undoubtedly require increased food availability far in excess of increased population. More land could be devoted to agriculture. The amount of land that could be used for crops is approximately three times the area currently under cultivation (Bongaarts, 1993), but much of this land is of inherently low productivity and is prone to environmental damage. Poor-quality land could be converted to agricultural use only at high economic and environmental cost. The distribution of potential cropland is highly skewed: some regions, such as sub-Saharan Africa and Latin America, are abundantly endowed, whereas others, such as large parts of Asia, are already near the limits of expansion. In temperate regions, agricultural and nonagricultural uses must compete for land, and substantial increases in agricultural land would occur only if the relative price of food were high enough. In most currently industrialized countries, land area under cultivation declined during the 20th century (Maddison, 1989, p. 21). Worldwide, total agricultural area (adjusted for multi-cropping) remained constant in the late 1980s.

Therefore, to ensure rising food production, yields must continue to increase. Inevitably, some of the yield-enhancing inputs that have been relied on in recent decades will become less effective; to keep output rising, others will have to compensate. Efficiency improvements are another possibility; Ali (1995) writes that a second green revolution in Asia might raise yields by a third, not by increasing inputs, but by easing the institutional and socioeconomic constraints that lead to inefficient use of existing inputs. Changes in diet could also play an important role (see *Box 5.1*).

In addition to improving productivity, ensuring food security will require growth in household income among the poor and an effective safety net of emergency assistance, entitlements, and special needs programs. Downing (1991) stresses that a range of strategies, many of them at the household level, are possible to reduce hunger, even under conditions of climate change. The question is whether agricultural policies will facilitate or impede such adjustments.

5.1.2   Impact of climate change

Climate change will affect agriculture in a number of ways (Parry, 1993; Downing, 1995; Reilly, 1996; Rosenzweig and Hillel, 1998).[6] Expected effects include greater frequency of climatic extremes, warming in high latitudes, possible changes in monsoon rainfall, and reduced soil water availability.[7] These must be set in the context of two direct effects of greater carbon dioxide (CO_2) concentration on

Box 5.1. Diet and the world food system.

Central to the world food situation is the composition of the food consumption basket (Goodland, 1997). Chen *et al.* (1990) calculated that present agricultural output, if uniformly distributed, could provide a minimal, but adequate, vegetarian diet to 5.5 billion people; Waggonner (1994) is more optimistic, estimating that current production could provide a vegetarian diet to a population of 10 billion. Assuming a diet in which 35% of calories are derived from animal products, as in more developed countries (MDCs), the same fixed resources could sustain only 2.5 billion people, less than half the current world population. Thus, how much growth in agricultural output is required to feed the planet will depend in large degree on the relative prices of animal and nonanimal protein. So far, consumers in less developed countries (LDCs) show every sign of following the well-worn path away from vegetable products and toward meat, poultry, and dairy products as their incomes grow (Ye and Taylor, 1995, for the case of China; Connor, 1994, for the "Americanization" of European dietary patterns). Yet others argue that there is substantial scope for changing the relationship between income and food demand. Even before allowing for the diversion of calories to feed use, close to one-third of global calorie consumption is wasted by inefficiencies in storage, handling, transportation, and food preparation.

   Are MDC food consumption habits wasteful and destabilizing? The answer is not necessarily simple. What is often called "the world food system" is a vast assemblage of institutions including future markets, transportation and storage facilities, credit markets, research and development institutions, information gathering and dissemination facilities, and the like. These would not exist without the scale made possible (indeed, necessary) by "wasteful" MDC dietary habits (Johnson, 1974). It could plausibly be argued that a "small is beautiful" world food system might err on the side of spotty markets, undercapitalized producers, and thin institutions.

plant growth: an enhanced rate of photosynthesis and greater efficiency of water use by plants. One of the major sources of uncertainty concerns hydrology: the linkage between hydrological events and climate patterns is unsatisfactory in general circulation models, or GCMs (Panagoulia, 1992; Liebscher, 1993, pp. 6–8; cf. Neméc, 1994 for the contrast between hydrological and climatological approaches to water problems). Hydrological events occur on smaller spatial and shorter temporal scales than those employed by GCMs. Although the uncertainty of regional climate change projections prevents us from drawing conclusions for specific regions, water impacts are expected to be greatest in regions that are already under stress, which include many arid or semiarid areas, particularly in less developed countries, or LDCs (Kaczmarek, 1996).

Global versus Regional Impacts

Given the range of effects and agricultural regimes, the net impact of global climate change on world agricultural production is inherently uncertain (Reilly, 1994). The first comprehensive assessment report of the Intergovernmental Panel on Climate Change (IPCC) reported that, based on research available at that time (e.g., Parry, 1990), it was impossible to conclude whether global agricultural potential would increase or decrease as a result of global climate change, but that prevailing regional patterns of production and trade would likely shift. The most recent IPCC assessment report (Reilly, 1996) concluded that global production probably could be maintained in the face of a changing climate, but confirmed and emphasized the earlier finding that regional impacts could be significant.

The view has long been held (e.g., many references in Kellogg and Schware, 1981; cf. Rosenberg, 1992, pp. 392–394, for a review of studies) that the general effect of global climate change will be a poleward shift of agroclimatic zones, with pronounced effects on the North American, Central Asian, and European grain complexes that are crucial to the world food system. For example, while climate change would probably permit expanded production of winter wheat in Canada, this would be at the expense of spring-sown crops, and the overall impact on North American grain production would be decidedly negative. Some areas currently under wheat cultivation would become semiarid (Rosenzweig, 1985), and some areas currently under corn cultivation would become suitable only for wheat. Changes in temperature and, more important, precipitation would lead to changes in runoff (McCabe and Hay, 1995; Nash and Gleick, 1991), probably increasing crop irrigation requirements. In contrast, in western Russia, where production is limited by short, cool growing seasons, climate change may improve yields. In Europe, the net effect of global climate change on agriculture would also probably be positive (Carter *et al.*, 1991). In northern, central, and upland areas where temperature is the main constraint on agricultural production, crop potential would rise; these gains would be partially offset by losses in southern and eastern regions where soil moisture, not temperature, is the limiting factor.

While early research concentrated on agriculture in temperate regions because of their leading role in the world food system, more recent work has extended the application of crop models to tropical regions. To the extent that agricultural adaptations are more difficult to make in low-income settings, the most vulnerable populations and thus the greatest human costs of climate change are likely to be in rural areas of LDCs (Reilly, 1996). Results of simulation studies are somewhat mixed, but tend mostly to indicate that impacts on LDC agriculture will be significant. Maytín *et al.* (1995) simulated the effects of different climate change scenarios on the yield of maize, a staple crop, at three sites in Venezuela. The middle-range results suggested that climate change might reduce maize yield by about 20% in a year of average precipitation and by more in a dry year. Mehrota and Mehrota (1995, p. 239) cite conflicting studies of the impact of global climate change on Indian rice and wheat yields; results differ depending on whether changing precipitation patterns and CO_2 fertilization are taken into account. In the cases of Mexico and southern Africa, respectively, Appendini and Liverman (1994) and Magadza (1994) conclude that climate change would reduce the availability of water for irrigation and significantly depress the production of staple crops. Depending on its effects on rainfall, a CO_2 doubling might either increase or decrease the availability of water for irrigation in Kenya (Kabubi *et al.*, 1995). Many studies on implications of climate change for Chinese agriculture are summarized by Smit and Yunlong (1996).

Because the poorest segments of the rural population depend on undesirable, ecologically unstable lands, the impact on the most disadvantaged members of society may be considerable. Thus, despite being relatively agnostic regarding the effects of climate change on global production, Parry and Rosenzweig (1993) expressed concern that global warming will lead to an increase in the number of people at risk of hunger. Downing (1991) has shown how a map of the African continent illustrating vulnerability to hunger corresponds closely to a map illustrating vulnerability to climate change.

Extreme Events

Extreme temperature or precipitation events can be defined in many ways; in the case of temperature, for example, they might reflect the number, length, or severity of heat waves. They also reflect the views of society and its ability to react. Drought, for example, is defined differently depending on the point of view: meteorological drought is defined as occurring when rainfall falls below a given level during a given time period, hydrologic drought occurs when streamflow falls below a given level, and agricultural drought occurs when crops are impaired as a result of insufficient moisture (Glantz, 1990). In regions prone to dryness, drought represents an intensification of problems that are encountered every year during the crop cycle, so a broad range of coping behaviors has emerged.

Extreme events undoubtedly are important climatic factors, and however defined, changes in their occurrence patterns will likely be among the most significant impacts of climate change. Changes could result from two general processes. First, changes in mean climate properties would affect the incidence of extremes. For example, summer temperatures are normally distributed, and global warming amounts to pulling the entire distribution slightly to the right. Thus, the probability mass in the extreme right tail will increase markedly, even if the variance of the distribution is unchanged.[8] Runs of consecutive summer "degree days" (Mearns *et al.*, 1984) would become more common. Second, changes in variability would change patterns of extreme events directly and could arise from shifts in storm tracks or changes in large-scale phenomena like El Niño events. Because they depend on smaller-scale features of the climate system, changes in variability are more difficult to project in GCMs, and confidence in these exercises is low. However, although in principle changes in variability could have a greater effect on the probability of extreme events than changes in mean climate properties (Katz and Brown, 1992, 1994), simulations suggest that, at least in the case of temperature, the effect of changes in the mean is dominant (Cao *et al.*, 1992).

While projections of changes in the occurrence of extreme weather events are uncertain (see Chapter 1), a few tentative general conclusions have been drawn (Kattenberg *et al.*, 1996): a warming climate is likely to lead to an increase in days with extremely high temperatures and a decrease in days with extremely low temperatures; some models suggest more intense precipitation events; and droughts may become more frequent or severe. Less confidence can be placed in studies at the regional scale; however, Cline (1992a) cites research indicating that an increase of 1.7 degrees Celsius (°C) in mean annual temperature would triple the frequency of heat waves in Des Moines, Iowa. Increased frequency of heat waves would not, however, occur farther north in North America or in any but the warmest part of southern Europe (Carter *et al.*, 1991). Matyasovsky *et al.* (1993) found that for eastern Nebraska, while growing season precipitation would increase (thus offsetting to some extent the effect of hotter weather), so would its variability. On the other hand, Parry and Carter (1990) point out that by reducing cold-summer damage, warmer conditions might also reduce yield variability, and that even if variability increases, proper selection of varieties could reduce exposure to risk.

As in the case of mean temperature, a relatively modest shift of mean annual runoff to the right or left can have much more severe impacts on the incidence of extreme events (Neméc, 1990; Chagnon, 1987). Random variations would not necessarily be statistically independent across river basins. For example, in years marked by the El Niño–Southern Oscillation phenomenon, streamflows worldwide are affected (Eagleson, 1994); indeed, changes in El Niño are correlated with a range of extreme climate events. Simulations with climate models have suggested that the variability of the critical South Asian summer monsoon might increase substantially with global climate change (Meehl and Washington, 1993).

If the variability of climatic conditions increases, much could be gained by improving the world agricultural system's capacity to balance out bad years against good (Rogoff and Rawlins, 1987). Much of this improvement could be realized by investment in agricultural infrastructure. On the other hand, when it is costly to reverse investment decisions (a reasonable assumption in the case of infrastructure), the otherwise simple relationship between variability and optimal level of investment becomes complicated (Fisher and Rubio, 1997).

CO_2 Fertilization

Complicating the outlook is the fertilizing effect of CO_2. An increased atmospheric concentration of CO_2 enhances photosynthesis, increases the efficiency of water uptake, and thus, all else being equal, raises yields (Rosenberg, 1982; Smil, 1990, pp. 11–12, for references). Experiments under controlled conditions have led some to argue that the beneficial effects on agriculture of rising CO_2 concentrations will more than compensate for adverse climatic effects (Smil, 1990, pp. 11–13, for references). While controlled experiments may be a poor guide to what will happen under actual growing conditions, where CO_2 fertilization must be combined with changes in temperature, soil moisture, and so on, research has tended to support the view that the beneficial effects of higher CO_2 levels will also be observed under field conditions (Rosenberg, 1992, pp. 388–390, for references). The major food crops that benefit most from increased CO_2 concentration are wheat, rice, and soybeans; those that benefit least are maize, sorghum, sugarcane, and millet. The first group comprises about four-fifths of world food production, so the net effect of CO_2 fertilization is positive. However, some regions would benefit much more than others. In assessments of the impact of global climate change on US agriculture (Adams *et al.*, 1995) and on maize yield in the European Community (Wolf and Van Diepen, 1995), CO_2 fertilization was found to be a decisive variable.

Adaptation

The ability of farmers to react to climatic adversity (Rosenberg, 1992, pp. 396–400) is substantial and has increased over time.[9] At the farm level, adaptive mechanisms include adjustments in planting and harvesting dates; changes in tillage practices, crop varieties, species, and rotations; fertilizer, herbicide, and pesticide applications; improvements in irrigation efficiency; and installation of new irrigation facilities (Parry and Carter, 1990; Reilly, 1996).[10] At the systemic level, changes in agricultural marketing, transportation, finance, national farm policies, and international agricultural agreements can all be invoked to mitigate the effects of global climate change. Mount (1994) divided adaptation into three stages: determination that new cultivars, crops, and management practices will work better under the new climatic conditions; adoption of the new practices by farmers; and market response to the ensuing changes in supply. At each step there is uncertainty and public policy can either enhance or thwart the adaptation process.

In part because of differing interpretations of the adaptability of the food system, the same climatic and agronomic scenario can give rise to widely differing interpretations. When integrated into economic analyses, agronomic results from Adams *et al.* (1990) pertaining to US agriculture gave rise to optimistic predictions by Nordhaus (1993) but to much more pessimistic ones by Cline (1992a, 1992b). Whether or not to include adaptations mediated through the price system is a crucial choice. For example, in a study that allowed price adjustments and international trade, Kane *et al.* (1992) found considerably smaller total welfare losses for the United States than did Adams *et al.* (1990), whose otherwise similar study did not include such paths of adjustment. When prices are allowed to play their role in allocating resources, changes in the distribution of welfare are more important than changes in total welfare. In the more pessimistic of their scenarios, Kane *et al.* (1992) found a net global welfare loss of only 0.5% of world gross domestic product (GDP); however, they projected major income transfers from food consumers to producers within countries and from grain importers to grain exporters internationally. Love and Shewliakova (1998) contrasted the situation in the United States, where policies and institutions have enhanced farmers' capacity to adjust to climate change, with the situation in LDCs, where the reverse has been true.

Even if losses in agricultural output are avoided, coping strategies will require the diversion of scarce resources from other worthwhile applications; thus, global climate change may cause the relative price of food to rise in order to attract these resources. If so, food-producing regions within countries and food-exporting countries will reap economic gains at the expense of food-importing regions and countries (Adams *et al.*, 1995). Neoclassical equilibrium models, which essentially solve for the agricultural prices necessary to clear world food markets, suggest that the magnitude of required price shifts will be modest, but the essence of such models is smooth substitution processes.[11] Rigidities – for example, agricultural protectionism – would increase the required price changes and thus magnify distributional impacts.

The role of variability should also be taken into account. Dalton (1997) found that including climate variability in an analysis of climate change impacts on agriculture significantly raised damage estimates.

5.1.3 Population and agriculture

Available research thus suggests that global climate change may raise additional challenges to an already challenged world food system. If that system would function better under conditions of slower population growth, then it stands to reason that climate change would increase the returns to policies designed to accelerate demographic transition. In this section, we look at the major inputs in agricultural production (land, soil, water, and technology) and in each case examine the role played by population.

Land

The debate over population and land, as well as other agricultural resources, mirrors the broader debate over population, natural resources, and the environment (see Chapter 3). The two polar positions are those of Malthus and Boserup. Malthus argued that, over the long run, population and agricultural resources remain in an equilibrium determined by the technology of food production and the minimum living standard. Because Malthus assumed that the agricultural resource base was fixed and made no allowance for technological progress, his conclusions were gloomy. Boserup (1965, 1981), however, argued that increasing population pressure itself induces technological change, leading to a more intensive use of land.[12] A refinement of the Boserup model is Hayami and Ruttan's (1971, 1987; Ruttan and Hayami, 1991) "induced-innovation" model, which draws a distinction between constraints imposed by an inelastic supply of labor and those imposed by an inelastic supply of land. The first, historically characteristic of more developed countries (MDCs), are offset by advances in mechanical technologies; the second, typical of LDCs, are offset by biological technology.

Boserup argued that opportunities for land extensification, that is, rural–rural migration, are exhausted before intensification commences. This might be called Ricardian adjustment, after the economist David Ricardo, who described an economy in which the very best land is cultivated first, then the next best, and so on until all arable land has been brought under the plow. Only when no more land is available do further increases in population density lead to substitution of more labor-intensive farming practices and modernization (mechanization, application of fertilizers and pesticides, etc.). The empirical evidence, however, is that land intensification sometimes occurs before exhaustion of land-extensification opportunities.[13] Pingali and Binswanger (1987) argue that medium- and large-scale public infrastructure investments, which facilitate land intensification at the farm level, become essential long before all potentially cultivable land is brought under the plow. They distinguish between farmer-based innovations and science- and technology-based innovations. Farmer-based innovations include changes in land use, land investments, organic fertilizer use, and the evolution of tool systems. Induced effects of population on these forms of innovation occur mainly at the household level. If high fertility worsens the vicious circle described in Chapter 3, then it stands to reason that it might impede farmer-based innovation. Science- and technology-based innovations include development of agricultural industry, science-based induced technological change, and development of agricultural research institutions. Induced effects of population on these forms of innovation are macroeconomic and have to do with the overall impact of population growth on economic development (including economies of scale in infrastructure provision).

The authors argue that farmer-based innovations can only support slow-growing populations. To accommodate rapid population growth, science- and technology-based innovations are necessary as well. If this is true, then the crucial question is whether rapid population growth impedes overall economic growth.

Complicating the picture is the fact that the distribution of benefits from technological change is affected by policy interventions, such as export taxes and food consumer subsidies (Alston *et al.*, 1988, for the case of industrialized countries; Anania and McCalla, 1995, for the case of developing countries). Some of these interventions are effectively endogenous; for example, rapid urbanization combined with failure of the agricultural sector to develop may lead almost inevitably to some form of consumer food subsidy. If population growth encourages policies that reduce the benefits of technological progress to those who must implement it (in this case, by forcing policymakers to subsidize rapidly growing urban populations at the expense of farmers), it would pose a serious obstacle to coping with demographic increase. The link between population and technological progress mediated by distributional issues may be critical and is entirely unstudied.

Land intensification can be either a virtuous or a vicious process, depending on the context. Properly implemented, intensification maximizes the productivity of existing land resources without impairing their long-term viability. Research in the Machakos district in Kenya (Tiffen and Mortimore, 1992; Tiffen *et al.*, 1994) and in semiarid northern Nigeria (Mortimore, 1993) has described how population pressure gave rise to farm-level innovations, elicited policy responses in the form of agricultural extension services, stimulated the growth of nonagricultural activities and off-farm employment, etc.[14] Farmers in the Kabala district of Uganda also coped successfully with rapid population growth (Lindblade *et al.*, 1998). Further evidence that demographic increase can stimulate productivity growth comes from the unlikely case of Bangladesh, where high-density rural districts had higher agricultural wages than low-density ones and, over time, the rate of population growth was positively correlated with the rate of change of real agricultural wages (Boyce, 1989).

However, intensification can also be a vicious, destructive process. The key task facing researchers is to identify the types of institutions that promote healthy responses to environmental stress. This is the focus of work by Bilsborrow (1987) and many collaborators (Bilsborrow and DeLargy, 1991; Bilsborrow and Okoth-Ogendo, 1992; Bilsborrow and Geores, 1994). The intervention of supportive, responsive institutions, especially in the public sector (Pingali, 1990), is crucial. Under adverse sociopolitical and institutional conditions, the result of population pressure will not be a virtuous Boserupian intensification, but either predatory cultivation of existing land resources or a destructive Ricardian extensification in which inequitable access to land causes poor families to crowd onto increasingly marginal land whose low productivity in turn exacerbates the problem of poverty (Heath and Binswanger, 1996). Lele and Stone (1989) document such outcomes in six sub-Saharan African countries. Cleaver and Schreiber (1993; cf. Kelley and McGreevey, 1994) describe an African population–agriculture–environment nexus that might be called "Boserup-in-reverse": a destructive cycle of rapid population growth, agricultural stagnation, and environmental degradation. The conclusion is that "policy-led intensification" is needed to supplement the demand-driven autonomous intensification described by Boserup. In Rwanda, characterized by rapid population growth and land degradation, Clay *et al.* (1998) identified four factors that promoted virtuous household-level intensification: secure land tenure, off-farm income, favorable profitability conditions for cash crops, and public investments in extension and roads. Case studies from Asia and Central America have illustrated the vicious combination of population pressure, poverty, and inequitable access to land that forces cultivators onto marginal lands (Cruz *et al.*, 1992), and the United Nations Secretariat (1991, p. 62) has concluded that there is reason to question the relevance of the Boserup model for many LDCs. Bilsborrow and DeLargy (1991, p. 144), having noted that the policy literature is filled with neoclassical policy recommendations for alleviating environmental deterioration, dismiss them as "dreamers' wish lists."

Agricultural Extensification and Deforestation

In the Tropics, trends in agricultural land use are closely tied to the problem of deforestation, which, as described in Chapter 1, produces nearly 40% of total LDC emissions of GHGs. Deforestation is a complex phenomenon, and its causes differ by region and have evolved over time (Rudel and Roper, 1996). One of the principal causes of deforestation is a process of agricultural extensification in which impoverished, landless cultivators encroach upon open, mostly secondary forest land (United Nations Secretariat, 1991, 1994). Sometimes governments promote extensification in order to relieve pressure on land in densely inhabited areas; other times it arises spontaneously from the vicious combination of poverty and the common-property, open-access aspect of the lands involved. Usually, landless cultivators, having no access to more productive land, are forced into "predatory" cultivation as an activity of last resort. Agricultural extensification of this type through the expansion of traditional shifting cultivation is not a result of demographic growth in the indigenous rural population, but of an increase in the number of households practicing shifting cultivation due to the arrival of new settlers (Dove, 1993, for the case of Borneo; Pfaff, 1999, for the case of the Brazilian Amazon).

The consequences have been staggering. Roughly half the original area of tropical forests has been lost to human activities, principally in 14 "main front" areas of deforestation containing about one-quarter of remaining forest cover (Myers, 1993a). The annual loss of tropical forest area, estimated by the Food and Agriculture Organization in 1980 to be 11 million hectares per year, was subsequently revised upward to nearly 20 million hectares per year (United Nations Secretariat, 1994, p. 27).

Many studies (United Nations Secretariat, 1991, p. 55, for references) have argued that poverty and inequitable access to good land are the root causes of extensification (Myers, 1993a, p. 12, for references; Bilsborrow, 1994, for a review of studies from the Philippines, Guatemala, Thailand, and Sudan). The nature of social institutions that allocate access to forests is crucial, and these vary widely. For example, when the owner does not restrict access, those who use the commons may fear that conservation on their part will encourage the owner to reestablish control over the resource (Azhar, 1993). The nature of off-farm rural labor markets is an especially important contextual variable (Bluffstone, 1995): poor opportunities elsewhere increase the intensity of farming and exploitation of non-timber forest products (Dufournaud *et al.*, 1995) and encourage households to fell trees for profit.

The role of population growth in extensification-driven deforestation is difficult to pin down. National rates of tropical deforestation have long been known to be correlated with the rate of population growth (Allen and Barnes, 1985; Mather, 1989; cf. Bilsborrow, 1994, pp. 126–128, for a review of studies), but results are highly sensitive to outliers (Bilsborrow, 1992; Bilsborrow and Geores, 1994) and it is difficult to infer any causality from this statistical association.[15] Rates of population growth in local areas undergoing extensification-related in-migration can be extremely high – sometimes in the double digits (Malingreau and Tucker, 1988) – and it is nearly impossible to explain such massive population redistribution in terms of national demographic factors. Thus, the national rate of population growth is a poor indicator or warning signal of deforestation (Myers, 1993a). Based on a meta-analysis of existing studies (i.e., statistical analysis of research results rather than raw data), Bilsborrow (1992) concluded that, while reducing population growth tends to ameliorate deforestation, its effect is modest because the process is dependent on many conditioning variables.

After agriculture, timber harvesting is the second most important cause of deforestation. Commercial logging has a greater effect than would be concluded based on the relatively modest forest area that it affects directly, because commercial development "leads the way" for cultivators (Walker, 1987, for the case of Borneo and Peninsular Malaysia). Some researchers (Vincent, 1992; Jonish, 1992, for the case of Malaysia) blame aggressive timber harvests on ill-conceived government policies designed to maximize near-term benefits. On the other hand, the empirical research base that isolates and analyzes the effects of government policies is still small, and some of the claims made on the basis of casual observation are at variance with results from general equilibrium economic models (Deacon, 1995). It has also been argued (Andersen, 1996) that while government policies, especially policies affecting access (such as road construction), may have a large impact on deforestation initially, local market forces eventually become the determining factors.

In semiarid regions, timber harvesting is often informal, due mainly to household consumption of biomass fuel. Despite expressing reservations about the magnitude of the problem, the United Nations Secretariat (1991, p. 57) wrote that the contribution of population to desertification through its effect on demand for fuelwood was "fairly clear." However, research cited in Chapter 4 concerning household-level economies of scale in energy consumption suggests that the relationship between population, households, and biomass fuel consumption is not simple. Second- and higher-order effects may also be significant; for instance, policies to discourage formal sector timber extraction may actually increase the rate of deforestation by shifting resources to the informal sector or to agriculture (Persson and Munasinghe, 1995).

Soil

Estimates of yield losses due to soil erosion in the major grain-growing temperate regions are modest (Crosson, 1992, for studies on the US case).[16] Because research has been concentrated in temperate regions, estimates of the impact of soil erosion and land degradation in LDCs tend to be qualitative at the local level (Dregne, 1990) and are little more than informed guesses at the national level (Colaccio *et al.*, 1989; Dregne, 1988). However, these estimates involve tremendous land areas (de Haen, 1991) and, because agricultural regimes differ, the finding that economic losses to soil erosion in temperate regions are small cannot necessarily be extended to tropical regions.[17] Working with an economy-wide general equilibrium model, Alfsen *et al.* (1997) estimated that the total direct and indirect costs of soil degradation in Ghana reduced the rate of economic growth by 0.6% per year. At the same time, semiarid lands are more resilient than once thought (Nelson, 1988), and there is recognition that the "desertification" crisis was overblown, at least on the global scale (Bie, 1990; cf. Sivakumar, 1992, for uncertainty regarding the long-term trend in rainfall in the Sahel region).

Population growth may encourage modes of production that have particularly negative consequences for soil. A vicious cycle of destructive intensification can be illustrated as follows. Consider an isolated, impoverished farming community cultivating a fixed land base that begins to experience diminishing soil productivity. In the near term, farmers meet an essentially fixed demand for food by adopting more labor-intensive methods of cultivation. The lower average productivity of labor implies a lower agricultural wage rate. Workers do not respond to the lower wage by seeking off-farm employment, because none is available; similarly, there is no possibility of out-migration. Only in the long run, when off-farm opportunities have developed or when resources have been mobilized to enable farmers to adopt land-saving methods of cultivation, is the basic problem – too much labor on too little land – addressed. Up to this point, it worsens and ecological damage results; the pace at which this proceeds reflects, among other factors, the rate of population growth. Similarly, in locales where land shortage forces cultivators farther up erosion-prone upland slopes, the pace of extensification will vary directly with the rate of demographic growth. These pessimistic scenarios are, however, conditional on failure of neoclassical substitution mechanisms, which is likely in some settings and unlikely in others. What is required for a virtuous response is a social, institutional, and economic context that enables, rather than impedes, the ability of households to respond to environmental stress (Blaikie, 1992), as in the Machakos experience described above.

Water

Fresh water is distributed unevenly around the world. Some regions, such as Scandinavia, have an embarrassment of hydrological riches, while others, such as the Sahelian belt in Africa, are experiencing worsening hydrological scarcity. The world's distribution of population is severely mismatched with the distribution of fresh water (Biswas, 1992; Gleick, 1993), and many growing populations are putting increasing pressure on the total volume of freshwater runoff.

Economic scarcity of fresh water comes about from the combination of supply and demand forces, which are mediated by human interventions in the storage, conveyance, and distribution of water. These interventions convert hydrological supply (stable runoff plus water stored in aquifers) into effective supply (the right kind of water in the right place at the right time).[18] Scarcity can therefore result from several causes (Falkenmark, 1994, pp. 109–110). Under some hydrological and climatic conditions, water evaporates before it reaches rivers or aquifers, resulting in a short growing season. Anthropogenic changes in the landscape, such as desertification and deforestation, give rise to unstable runoff and soil erosion, which pollutes streams, rivers, lakes, and reservoirs (Smil, 1993, for the case of China). Finally, water scarcity can arise from direct population pressure on available supplies and/or from gross inefficiencies in water use. Projecting water scarcity is a difficult exercise in which global factors (in the form of climate change), regional factors (in the form of land-use change), and changes at the river-basin level (in the form of water resource management) must be taken into account (Conway *et al.*, 1996).

Agriculture, and thus irrigation, plays the dominant role in water demand. Over 70% of global freshwater use is for irrigation, although this varies with local circumstances (World Resources Institute data cited by Falkenmark and Widstrand, 1992, p. 14; Biswas, 1992, 1994a).[19] Expansion of irrigation can induce changes in river delta ecosystems with resulting impacts on local climate (Smith, 1994, for the case of the Aral Sea basin; Thompson and Hollis, 1995, for the impact of water resource schemes on wetlands). After agriculture, the next largest claimant is industry (23% of global water use), with domestic use coming in a distinct third (8% of global water use). Thus, despite varying local circumstances, and while it is easy to condemn wasteful domestic water consumption in MDCs or the growth of megacities in LDCs, the key to ensuring future water supplies – including satisfying the needs of rapidly growing urban populations – lies in improving the efficiency with which the resource is used in agriculture.[20]

While there are further opportunities for development of irrigation, supply constraints and the rising value of competing uses are increasingly felt (Falkenmark and Widstrand, 1992). The number of large dams completed annually has steadily declined and many aquifers are being overexploited (Kavalanekar *et al.*, 1992, for the case of the Mehsana aquifer in Gujarat). The rate of growth of irrigated area has declined since the early 1970s in all regions except Africa (Biswas, 1992). Easy opportunities for irrigation have, for the most part, already been exploited; moreover, substantial amounts of land (precise estimates are not available) currently under irrigation are experiencing or are threatened by salinization due to failure to invest in adequate complementary drainage systems (de Haen, 1991).

By concentrating productive resources in agriculture, population growth raises demand for water in the sector that is the least water efficient. Whether this increased demand is or will actually be met is another question: Biswas (1994b) has written in the Asian case that, because priority is either explicitly or implicitly given to domestic uses of water, the irrigation sector will simply have to learn to do more with less. In a perfect neoclassical economic world, rapid population growth would offset itself by encouraging the rationalization of water use. Yet there is little disagreement that existing water resource management policies are inadequate. Serageldin (1995) has summarized the problems as follows: first, most countries refuse to treat water as an economic good, leading to gross waste; second, water resource decision making has been overcentralized in government, leading to failure to take the views and preferences of local stakeholders into account; third, water management is fragmented among agencies, leading to failure to deal with conflicts and complementarities between sectors; fourth, health and environmental concerns are underappreciated. In a study of China, India, Pakistan, and South Korea, Kaczmarek *et al.* (1997) found that, while climate change gave cause for concern, population growth and, most fundamentally, poor water management were the main problems facing policymakers.

Studies in both MDCs and LDCs have repeatedly identified situations in which water that could be used more productively for urban domestic and industrial purposes is used instead for agriculture because of subsidies in the form of subsidized capital costs for irrigation projects and below-marginal-cost pricing of water delivered to farmers (World Bank, 1992). In LDCs, the delivered price of water seldom pays more than a fraction of actual supply costs; thus, it is not surprising that irrigation is the most inefficient user of water supplies, with waste sometimes exceeding 70% of total water diverted. Worldwide, only 45% of water diverted or extracted is effectively used by the crop; 25% is wasted in field application losses and 15% apiece in farm distribution losses and irrigation system losses (FAO, 1994, cited by Serageldin, 1995). Beaumont (1994) estimates that less than 200 hectares of irrigated Egyptian farmland consume water equivalent to the domestic use of a city of 1 million (at prevailing Egyptian per capita domestic water consumption levels). These 200 hectares produce less than 1,000 tons of wheat, which could be bought on the world market for less than US$200,000. The opportunity costs can easily be appreciated. In the case of the Middle East, Dinar and Wolf (1994) identify significant efficiency gains to trade in water and conclude that the main impediments are neither conceptual nor technical, but political. In many regions, charging water prices that reflect the scarcity value of the resource would price irrigated crops out of the world market. Under these circumstances, policymakers must choose between maintaining an unsustainable natural resource policy and forgoing production of the crops in question, which may mean depending on imports. Importing food is often a bitter pill for policymakers to swallow, even when comparative advantage dictates that it is in the economic interests of a country to do so.

Thus, charging farmers the scarcity value of irrigation water may not always be politically feasible. Moreover, there is a hard core of circumstances under which the pricing of water in conventional terms is hydrologically impractical and conventional approaches have failed. Estimates of the price elasticity of demand for irrigation water in the western United States vary widely (Malla and Gopalakrishan, 1995, pp. 235–236). Efficient outcomes mediated through the price system may also be inequitable (du Bois, 1994). When water comprises a substantial portion of wealth, market forces exacerbate income inequality: the rich are able to purchase more water, thus augmenting their wealth and purchasing power. Managed allocations, on the other hand, tend to lag behind social and economic change (for example, continuing to provide subsidized water to an uncompetitive agricultural sector) due to institutional and political inertia.

Perry (1995) identified three prerequisites for a successful irrigation scheme: assignment of water rights, existence of infrastructure capable of delivering the entitlements implied by water rights, and assignment of operational responsibilities. The principal challenge to irrigation in LDCs, he argues, is the reassignment of water rights, because it gives rise to large, instantaneous transfers of wealth. Such transfers notwithstanding, there are examples of decisive institutional changes in the water sector (González-Villareal and Garduño, 1994, for the case of Mexico).

Under such circumstances, the pertinent question is not whether population makes water more economically scarce – indeed it does, and this scarcity ought to be reflected in rising water prices. The question is whether population pressure makes it more difficult for authorities to deal with the equity and distributional consequences of pricing water competitively. Biswas (1998) is pessimistic about the readiness of decision makers to come to grips with water problems. If slowing population growth gives policymakers (and institutions generally) more breathing room in which to make decisions, then in a time when global climate change poses difficult allocational choices, a world of moderate population growth is bound to be preferable to one of rapid demographic increase.

Technology

A theme that emerges repeatedly in research is that expansion of the agricultural knowledge base is crucial. This was stated most explicitly by Crosson and Anderson (1994), who concluded that although potential supplies of land and water for irrigation are ample, the economic and environmental costs of mobilizing them are so high that neither one will make a substantial contribution to meeting world food demand in coming decades. Only increase in the knowledge resource will allow demand to be met, and production and application of knowledge on the requisite scale are likely to be costly endeavors. To get an idea of the demands placed on the knowledge base, note Binswanger's (1989) estimate that agricultural prices would have to increase 5–10 times to elicit a sufficient supply response in the absence of any improvements in knowledge. Because agricultural knowledge production is heavily farm based, the development of local capacity to identify problems and design locally appropriate responses has repeatedly been stressed. The sorts of conditions that encourage locally based invention, innovation, and technological progress in agriculture are broadly similar to those that will encourage virtuous adjustment to the challenges posed by climate change. These are the existence of appropriate property rights, the smooth and equitable functioning of institutions, and the alleviation of poverty and insecurity.

5.1.4 Conclusion

By placing growing demands on the world food system, population growth raises the stakes in global climate change. Given population trends, food availability will have to increase three- or fourfold by the middle of this century to achieve the goals of improved diet and enhanced food security. This implies that the 2% long-term rate of increase in world food supplies since the mid-1930s – an astonishingly rapid rate of increase by historical standards – will have to be maintained in coming decades. While climate change may be neutral at the global level, it is likely to make attaining necessary agricultural growth more difficult in the poor regions that need it most. Maintaining global agricultural production under conditions of climate change may require an increase in the relative price of food, which hurts the poor.

Population has impacts at the household- and macro-levels. At the household level, high fertility might impair the capacity for innovation and constructive agricultural intensification along lines described in Chapter 3. Even if it does not, research indicates that household-based innovation alone is insufficient to cope with rapid population growth. At the national (and global) level, the question is whether rapid population growth encourages or impedes the development of effective institutions – such as markets, research laboratories, extension services, transportation networks, and the like – to enhance agricultural growth. This is closely related to the broader question of whether population growth hinders general economic development and progress. As the distribution of food and agricultural resources is an important component of the overall distribution of wealth and income in LDCs, there is also the important question of whether slower population growth would make it easier for policymakers to make difficult allocational decisions.

On balance, the research described above suggests slower demographic growth would ease pressure on the world food system and make it more resilient to the stresses expected from climate change. On the other hand, the situation is far too complex and contingent on local circumstances and institutions to support any claim that reducing the rate of population growth is a key strategy in this area. There are too many other avenues of improvement to be pursued, including the reduction of gross inefficiencies in food production, storage, distribution, and consumption.

5.2 Health

Health, whether at the level of the individual or the population as a whole, is an outcome that integrates many inputs. Some of these are outside human control and some, such as social and individual responses to disease and perceptions of "good health," depend on human behavior. As an essential element of human capital, good health is crucial to development and interacts closely at the household level with factors such as poverty and fertility.

Climate change is likely to have both direct and indirect effects on health, and on balance impacts are expected to be adverse (McMichael, 1996). We review the broad outlines of the world health situation in terms of the mortality transition over the past several decades. We then review the potential impact of climate change on health and conclude by examining the role that slowing population growth, relative to other types of interventions, might play in improving the resilience of societies to climate impacts on health.

5.2.1   Current situation

As reviewed in Chapter 2, a fairly typical transition from high to low fertility can be discerned in the historical record. An equally evident mortality transition, typically preceding the fertility transition by several decades, can be described as well. In MDCs, this transition was marked first by increased resistance to disease as a result of improved living conditions, then by mortality declines due to improved sanitation, and then by declines due to medical advances such as vaccinations and antibiotics (McKeown, 1988). The effect of these developments was to concentrate mortality in older age groups, where degenerative, rather than infectious and parasitic, diseases predominate. The most recent phase of the mortality transition has been reduction in some of these causes of death, such as cardiovascular disease and cancer.[21] Despite advances in high-technology, high-cost medical interventions, mortality rates in MDCs are more closely associated with the equity of the income distribution than with income level itself (Wilkinson, 1994).

The transition has been different in LDCs, because modern medical technology became available before these countries had undergone improvements in living conditions. Thus the mortality decline that took 200 or more years to occur in the West has occurred in the space of 50 years in LDCs. One result has been a gap in the medical care available to the rich and the poor (World Bank, 1993). For example, in a country where rural areas often lack basic maternal and child health care facilities or sanitation systems, the capital city may possess a state-of-the-art center for cardiac surgery. While the urban–rural gap in health care is still the main concern, the gap between the urban middle class and poor is also a problem, as is that between elite workers with access to the public health system through membership in social security schemes and the rest of the population (McGreevey, 1990). Reforming health systems to eliminate inequities and improve the health of the poor is one of the most promising opportunities for combating poverty (World Bank, 1993).

Health transition in LDCs must be placed in the context of environmental conditions. In rural areas, agricultural intensification has led to the growing use of machinery, fertilizers, and pesticides, often under conditions considered unsafe in industrial countries (Painuly and Rev, 1998). More general agronomic practices may affect health as well; for example, mosquito breeding is affected by how often rice fields are drained.[22] Urban areas, on the other hand, are exposed to elevated levels of both traditional pollutants, such as particulates from fuelwood or charcoal burning, sulfur dioxide, and urban smog (Romieu *et al.*, 1990; Pope *et al.*, 1995; Alberini *et al.*, 1997), and new pollutants, such as toxic industrial chemicals. Rapid industrialization has not been accompanied by the development of institutions and regimes to ensure proper disposal of the harmful wastes generated by industries (Ludwig and Islam, 1992, for Bangladesh; Asante-Duah and Sam, 1995, for West Africa). Marquette (1995) has written of the urgent need for improved information-gathering and planning mechanisms to cope with environmental health hazards in rapidly industrializing countries and in the transition countries of Eastern Europe and the former Soviet Union, as well as of the need for a greater focus on the urban environment in a rapidly urbanizing world.

Several aspects of the world health situation stand apart from the standard "mortality transition" model. The primary deviation is the widespread increase in human immunodeficiency virus (HIV) infection and ensuing acquired immunodeficiency syndrome (AIDS) mortality. Studies discussed in Chapter 2 have concluded that the effect of AIDS mortality on population growth, while significant in some countries, is likely to be modest at the global level. On an individual and societal level, on the other hand, the AIDS epidemic is an epochal event, roughly equal in scale to the 1918 influenza pandemic. In Africa, the social effects of AIDS are exacerbated by the fact that the persons at highest risk are in their prime productive years and belong to skilled urban elites (Ainsworth and Over, 1994). Thus, while other diseases such as tuberculosis and malaria account for more deaths, the economic impact of AIDS is greater. In all regions, AIDS is placing stress on public health budgets and, more generally, increasing the share of total resources that must be devoted to health (for example, placing claims on savings of affected families, raising insurance premiums, etc.).

Other exceptions to the standard mortality transition model exist as well. Recent years have seen a resurgence of infectious diseases and the emergence of new viruses besides HIV (Morse, 1991). Examples include the emergence of new vector-borne viral infections in the Americas, the discovery of a new and spreading strain of cholera, the resurgence of diphtheria in Russia, and the emergence of various antibiotic-resistant microbes, including the malaria parasite (especially in Asia) and multidrug-resistant tuberculosis strains (among urban populations in the northeastern United States). Also well publicized has been rising mortality in the transition economies of Central and Eastern Europe (Feachem, 1994; Potrykowska and Clarke, 1995). Some of these exceptions, like the appearance of HIV, are classic "surprises"; others, such as the development of antibiotic-resistant bacterial strains, fall within the range of experience and could have been predicted by models that reflect the ecological complexity of disease.

5.2.2   Impact of climate change

During two centuries of urbanization and industrialization, public health hazards – infectious diseases, chemical pollutants, occupational exposures, and "affluent" risk-bearing behavior such as consumption of tobacco, alcohol, and processed foods – have emerged primarily at the local scale, reflecting local demography, culture, technology, and wealth. In contrast, climate change poses a new, supra-population scale of health hazard (McMichael, 1993, 1996; Last, 1993; Epstein, 1995).

Forecasting potential health impacts entails multiple uncertainties, probable nonlinearities, and, for many impacts, a long time horizon (McMichael and Martens, 1995). The first two characteristics pose challenges to prevailing modes of scientific data assembly and analysis, and subsequently to the process of scientific inference. The long time horizon suggests that much of the research will not be amenable to conventional hypothesis testing (certainly not within the time frame of relevance to contemporary decision making). In addition, predicted exposures may be qualitatively different from past experience, because they will reflect changes in a range of background variables, such as changes in nutrition and the distribution of disease-bearing vectors (ticks, mosquitoes, fleas, etc.). Because of the complexity of the ecological and geophysical disturbances attendant on climate change, and of the relationships between disease incidence, nutritional status, and morbidity/mortality outcomes (Murray, 1994), we should expect some surprise outcomes (Levins, 1995). Background conditions, such as per capita income, literacy, nutrition, adequacy of public health infrastructure, etc., are likely to be better in the future than they are today. However, as developments such as the recent emergence of multidrug-resistant tuberculosis remind us, monotonic improvements in health cannot be taken for granted.

Despite the inherent difficulty of projecting impacts, the bulk of research has concluded that the net impact of global warming on human health is likely to be negative (McMichael, 1996). Given the possibility for protective response, the health impacts of climate change in MDCs will probably be modest. In LDCs, on the other hand, where populations are more vulnerable (both socially and biologically) and where public health spending is limited by available resources, mortality and morbidity conditions are likely to worsen as a result of global warming. This is not to say that health conditions will be worse than they are today, because continued improvements in living standards are likely. However, health conditions are likely to be worse than they would have been in the absence of climate change.

Direct effects include increased morbidity and mortality due to a greater frequency of extreme events, particularly heat waves, storms, floods, and fires. Flooding gives rise to an entire class of public health problems (Gregg, 1989), including outbreaks of diarrheal disease when water sources are contaminated by fecal material, problems associated with runoff of toxic materials (Thurman *et al.*, 1991, 1992), and increases in some vector-borne infectious diseases (Cotton, 1993; Hederra, 1987). Even without flooding, increased rainfall and higher temperatures would also be favorable to the bacteria and protozoa that cause diarrheal diseases, including cholera.

Indirect effects, though more difficult to forecast than direct effects, are expected to be more significant in the long term. They include the following:

- Changes in infectious disease epidemiology, particularly increases in the spread and activity of a variety of vector-borne diseases.
- Disruption of freshwater supply, and perhaps sanitation, because of altered precipitation patterns and sea level rise.
- Extensive population movements, some involving distress as discussed in Section 5.3.
- Health impacts of social stresses associated with global environmental change and its impact on the economy.

In this section, we focus on two health impacts that are most likely to prove significant and that have been most intensively examined: the direct effects of increased frequency of thermal extremes and indirect effects mediated through alteration of the distribution of vectors that transmit certain infectious diseases. While the first is much better understood, the second is likely to involve greater impacts because of the greater size and vulnerability of the population exposed to heightened risk and the more limited resources available for response and adaptation.

Thermal Extremes

The main thermal hazards of global warming would result from increased frequency and severity of heat waves, often defined as five days in a row in which the daily ambient temperature exceeds the normal body temperature of 37°C (98.6°F). After several days, prolonged heat stress can lead to heat exhaustion, characterized by dizziness, weakness, and fatigue. More acutely, heat stroke, in which body temperature exceeds 41°C (106°F), may occur, often resulting in unconsciousness and sometimes death.

Healthy persons can cope with moderate increases in ambient environmental temperature and, within certain limits, physiological acclimatization develops after several days, minimizing heat stress. However, frail or ill individuals with lower physiological resilience do not adapt as well, making heat stress a greater health hazard to elderly persons and infants, and to persons suffering from cardiovascular or other disorders.

Higher summer temperatures and accompanying heat waves would increase rates of heat-related illness and death in both temperate and tropical regions, but reliable dose-response estimates exist only for the first region. The extremely hot summers experienced in Missouri during the early 1980s might be considered a historical analogue. Temperatures were 2–3°C higher than normal and heat-related deaths occurred at seven times their normal rate (Centers for Disease Control, 1989). The pattern of deaths from all causes in relation to daily summer temperature in New York City indicates a nonlinear relationship with a threshold of approximately 33°C, above which overall death rates increase sharply (Kalkenstein, 1993b). However, studies in the United States indicate that 20–40% of such heat-related deaths during heat waves represent a displacement or "early harvesting" effect, that is, they represent deaths that would have occurred within several weeks under normal temperature conditions. From studies done in "heat-sensitive" northern US cities – those where heat waves are infrequent – it has been estimated that, under climate scenarios accompanying a doubling of CO_2 concentration, heat-related deaths in such populations would increase by a factor of five or six (Kalkenstein and Smoyer, 1993). Cities with naturally warmer weather would be less affected, because their inhabitants are better adapted to prolonged bouts of summer heat.

Excess mortality related to heat waves is greatest in crowded urban environments, due to the "heat island effect," and in communities lacking air conditioning and proper ventilation. Before the advent of air conditioning in industrialized countries, excess mortality during heat waves was much more pronounced than it is at present. Data from heat waves of similar magnitude in Los Angeles in 1939, 1955, and 1963 indicate a sharp drop in excess mortality between the second and third episodes, probably due to the advent of residential air conditioning (Goldsmith, 1986).

Just as global warming is expected to increase heat-related mortality, it is expected to reduce wintertime mortality by reducing the extent and frequency of extreme cold weather (Kalkenstein, 1989, 1993a; Langford and Bentham, 1993). Excess winter mortality is predominantly due to influenza, other respiratory infections such as bronchitis and pneumonia, and coronary heart disease. Although the overall balance is difficult to quantify, it is likely that sensitivity of death rates to hotter summers will be greater than sensitivity to milder winters in most countries.

Higher summer temperatures would also stimulate the chemical reactions that give rise to photochemical smog. Bufalini *et al.* (1989) estimated that a 4°C increase in average summer temperature in the San Francisco Bay area would triple person-hours of exposure to concentrations of atmospheric ozone in excess of the current air quality standard. Exposure to such pollutants reduces lung function, thus increasing susceptibility to infection, heat stress, and chronic lung disease, with consequences for mortality and morbidity from a wide range of causes.

Vector-Borne Diseases

While morbidity and mortality related to thermal extremes are significant for sub-populations such as the aged, they are of minor (and declining) overall public health significance. Much more important will be the impact of climate change on infectious diseases. Most infectious disease transmission mechanisms are affected by climatic factors in one way or another (Bradley, 1993), and the link to climate is especially close in the case of vector-borne infections.[23] Most vector-borne diseases have at least three, and usually four, components (Longstreth, 1990): the infectious agent (e.g., to take the case of Rocky Mountain spotted fever, *Rickettsia rickettsii*), the vector (*Dermacentor variabilis*), an intermediate host (woodland birds and animals), and an ultimate host (humans). Changes in climate can affect any one of these components.

The maintenance of a vector-borne infectious disease agent requires an adequate population of the vector and favorable environmental conditions for both vector and parasite (Shope, 1991; Koopman *et al.*, 1991; Herrera-Basto *et al.*, 1992; Loevinsohn, 1994; Bouma *et al.*, 1994). Slight changes in climate can affect the viability and geographical distribution of vectors. To take one example, and probably the most important one, the *Anopheles* mosquito, which transmits malaria, does not survive easily where the mean winter temperature drops below approximately 15°C, and survives best where mean temperature is 20–30°C and humidity exceeds 60% (Martens *et al.*, 1995a, 1995b). Higher temperatures accelerate the developmental cycle of the malaria parasite (Burgos *et al.*, 1994). Further, the malaria parasite (*Plasmodium*) cannot survive below a critical temperature: around 14–16°C for *P. vivax* and 18–20°C for *P. falciparum*.

By improving conditions for both vector and parasite, unusually hot, wet weather in endemic areas can cause a marked increase in malaria incidence (Loevinsohn, 1994). A simple model that related annual average temperature and rainfall to malaria incidence predicted that incidence in Indonesia might increase 25% over 80 years as a result of climate change (Asian Development Bank, 1994). Another model employing the same approach for the world as a whole found that five different climate change scenarios gave rise to increases ranging between 7%

and 28% in the land area potentially affected by malaria (Martin and Lefebvre, 1995).

An epidemiological model has been developed (Martens *et al.*, 1994, 1995a, 1995b) in which development, feeding frequency, and longevity of the mosquito and the maturation period of the parasite within the mosquito are endogenous variables. The prevalence of public health control measures is exogenous; thus, the model's predictions refer only to potential alterations in the geographic range of transmission. Their significance for alterations in disease incidence must be interpreted in relation to local conditions and public health control measures. Model simulations based on an assumed increase of 3°C in global temperature by 2100, holding population size and public health responses fixed, indicate a doubling of potential malaria transmission in tropical regions and a 10-fold increase (from a very low base) in temperate regions. A core finding of the study was that, while potential malaria transmission would be increased in all currently endemic regions, the effect would be more severe in currently low-endemicity areas than in areas where prevalence is already elevated.[24] Included among areas that might become endemic are highland urban areas, such as Nairobi and Harare, or areas in the Andes and mountainous western parts of China that are currently situated above the "mosquito line." The vulnerability of newly affected populations would initially lead to high fatality rates because of their lack of natural immunity.

The reemergence of malaria in formerly malarial areas, such as parts of the southern United States, is unlikely. Longstreth (1990), interpreting results of model simulations by Haille (1989) as well as other research, wrote that only a rather unlikely combination of events could lead to the reemergence of malaria as a serious public health problem in the United States: a breakdown in the effectiveness of vector-control programs or establishment of a pesticide-resistant strain of *Anopheles* plus the emergence of a large infected human population left untreated for an extended period of time.

Malaria is only one of many vector-borne diseases whose distribution may extend and whose intensity may be increased by warmer, more humid conditions (Rogers and Packer, 1993). Others include trypanosomiasis (African sleeping sickness and American Chagas disease), filariasis (elephantiasis), onchocerciasis (river blindness), schistosomiasis ("bilharzia"), hookworm, Guinea worm, and various tapeworms. Vector-borne viral infections such as dengue fever, yellow fever, and rodent-borne hantavirus are also affected by temperature and surface water distribution (McMichael *et al.*, 1996). Dobson and Carper (1992) have put it simply: parasites and disease will do well on a warmer Earth. Against this must be set the improvements in living conditions and advancements in medical technology and public health institutions that will occur during this century.

5.2.3   Population and health

Population is related to health in several ways.

- As studies reviewed in Chapter 3 have documented, settings characterized by low income and high fertility are likely also to be characterized by poor household-level health. At least some of this association represents an independent causal effect attributable to high fertility. The health of women and children is definitely impaired by closely spaced births, which in turn are associated with high completed fertility. A link which may prove especially significant in the context of climate change is that between maternal and child health and susceptibility to malaria.
- In rapidly growing populations, scarce public health resources are strained. However, government health expenditures per capita are largely uncorrelated with health outcomes such as infant and child mortality and life expectancy. "Dilution" effects are a far less important policy problem than inefficiencies and inequities in the distribution of available health and education resources – for example, the bias in favor of curative hospital care at the expense of primary health care, the bias in favor of boys at the expense of girls, and so on (Jiminez, 1989).
- In low-fertility societies, population aging places special demands on the health system. It is cause for concern that health systems will have to cope with any stresses arising from global climate change at the same time that they cope with the challenges posed by population aging. While the social challenge of population aging is currently most apparent in MDCs, LDCs will experience aging much more rapidly because of the speed of fertility decline.

Population is also linked to health through poverty. Poverty is inimical to health, and poor health is, in turn, a form of impoverishment. While it has proved difficult to formally establish a causal relationship between poor health and low income, the weight of evidence increasingly supports the proposition that poor health gives rise to adverse economic outcomes (Strauss and Thomas, 1998). If policies affecting population can reduce poverty, then they will have important spillover effects on the health of the population and vice versa if they can improve health.

5.2.4   Conclusion

The evidence suggests that reducing high fertility would have some beneficial effects on health at the household level. Some policies that tend to lower fertility, such as maternal and child health programs or programs to promote the education of girls, may also have beneficial impacts on health independent of their impacts on fertility. Healthier societies with stronger health institutions are more likely to be resilient to the impacts of climate change.

However, simply slowing the rate of population growth is unlikely, in and of itself, to be a very effective policy for improving health for a number of reasons. First, it cannot substitute for a more equitable distribution of available health resources. Second, there is the danger that the population aging that follows from slower population growth will require additional resources without improving the overall health of the population; indeed, it may sharpen the conflict between expenditure for curative care, which benefits mostly aged patients and the middle and upper classes, and the demand for preventive care, which benefits mostly young persons and poor households. As we discussed in Chapter 2, by the middle of this century, the proportion of the LDC population that is elderly (over 60) will approximately equal the current elderly share of the MDC population. Policies that have the impact of lowering fertility in LDCs will significantly raise the elderly share, with accompanying stress on health care institutions (World Bank, 1994).

Third, a wide range of more direct measures to improve resilience to the health impacts of climate change are available. For example, in the case of malaria these include surveillance, improved treatment, open-water management, and applications of pesticides (although the acceleration of the life cycles of parasites allows quicker development of resistant strains). Better and larger-scale public health monitoring systems are also needed (Haines *et al*, 1993). Interventions need not be designed to directly reduce incidence of the disease, but rather to reduce its impact on the community. For example, where malaria eradication is infeasible, interventions designed to improve the nutritional status of an undernourished population may be more appropriate than vector control interventions. As in the case of agriculture, the importance of local capacity and institutions must be stressed.

5.3   Environmental Security

According to one school of thought, the pressure of impoverished populations against renewable natural resources is becoming one of the leading causes of population displacement, internal conflict between different ethnic and interest groups, and, ultimately, international conflict (Homer-Dixon, 1991; Homer-Dixon *et al.*, 1993). If present trends continue, it will be one of the major international security problems of this century. Bonneaux (1994) and Engelman and LeRoy (1995, p. 30) all but endorse the hypothesis that recent violent internal conflicts in Rwanda, Somalia, Yemen, and Haiti are due in significant degree to rapid population growth. Much of the focus is on Africa, where drought and desertification have already taken a heavy toll; South Asia, particularly sensitive border areas such as the territory between India and Bangladesh, is another area of concern.

In these scenarios, population growth is assumed to be a contributing factor because it exacerbates stresses on natural resource systems; therefore, slowing population growth could ameliorate the problem by easing that stress. We implicitly addressed this possibility earlier in the chapter in our consideration of population and agriculture, land, soil, and water, concluding that, on balance, slowing population growth would improve the ability of societies to adapt to stresses placed on these resources, although it is not likely to be a key strategy since many more direct measures are available. In this section, we take up a different, but related question: Is climate change likely to significantly worsen environmental security?

The environmental security debate has been characterized by two fundamentally different constructions of the problem. On the one hand, environmental security has military and political implications (Elliott, 1996). If West European states are at risk of contamination from nuclear power plants located in Eastern Europe and the newly independent states, then in a sense their national security is threatened by environmental risks. On the other hand, environmental security also can be constructed as reasonably secure access to a dependable flow of environmental services. In this framing, environmental security is an issue mainly for the rural poor in developing countries. A synthesis of the two views is that, when access to environmental resources in poor regions is impaired, these regions become political and military tinderboxes. In the following pages, we concentrate on this synthetic view.

5.3.1   Current situation

Perhaps no phrase has brought population–environment–development linkages home to policymakers more forcefully than "environmental refugees," evoking as it does images of human misery and social chaos (Ramlogan, 1996). Defining "environmental refugees," however, is problematic. For example, El-Hinnawi (1985) included "all displaced persons … having been forced to leave their original habitat (or having left voluntarily) to protect themselves from harm and/or seek a better quality of life." Myers (1994b) offered a similarly broad definition: "People who can no longer earn a secure livelihood in their erstwhile homelands because of drought, soil erosion, desertification, and other environmental problems." Yet granting all people displaced under such conditions "refugee" status may overstate the case. Population displacements associated with environmental distress occur along a continuum, with refugee flight at one end and normal migration at the other. Refugees, in common usage, flee involuntarily and in haste; they are powerless and vulnerable in their new place of residence. While not in strict accordance with international law, the term "environmental refugee" might reasonably be applied to persons displaced by a flood that in turn was exacerbated by environmental deterioration.[25] Migrants, in contrast, move of their own volition in response to a

combination of disagreeable conditions (push factors) and in anticipation of a better life (pull factors); once installed in their new residence, they typically relate to
the host society from a much stronger position than do refugees.[26] The members
of a household that relocates after months of struggling to cope with worsening environmental conditions are probably better described as "environmental migrants"
than as "environmental refugees."

Some authors have argued that substantial numbers of migrants in LDCs represent households that have been displaced by deteriorating environmental conditions
and forecast that the situation is bound to worsen. Westing (1992, 1994) estimated
that there were 10–15 million displaced persons, mostly in Africa, and that the
number was increasing by some 3 million per year. Arguing that there has been
no discernible rise in warfare or persecution, he assigns the increase to environmental deterioration. Myers (1994b) calculated that, by the year 2050, the number
of environmental refugees may rise to 150 million; some 1.5% of projected world
population.

Some researchers have also linked impairment of renewable natural resource
systems and subsequent migration to acute internal and international conflict
(Westing, 1986; Renner, 1989). By far the largest body of research dealing with
conflict over natural resources has dealt with water (Falkenmark, 1989; Starr, 1991;
Gleick, 1993, 1994; Anderson, 1992), where conclusions have varied widely. For
example, Lonergan and Kavanaugh (1991, p. 281) forecast "desperate competition
and conflict" in the Middle East in the near future, even discounting the long-term
effects of climate change, while Beaumont (1994) views water wars in the region as
a "myth." Lying behind such divergent points of view is the fact that water scarcity
is just one of many factors that can lead to conflict. Gleick (1992; cited in Smith,
1995) identified four factors that determine a country's vulnerability to disruption
of shared water resource supplies: scarcity of water, extent to which water is shared,
degree of dependence on shared water, and the relative power of riparian states. In
judging whether disruption is likely to lead to conflict, additional factors such as
perceived interests and internal and external power relations must be considered.
Furthermore, if low-cost internal adjustment mechanisms are available, states need
not resort to more drastic measures. Beaumont's (1994) optimistic outlook on the
Middle East stresses the tremendous inefficiency with which water is employed in
much of the region.

When the costs of internal adjustment are too high, recourse to mechanisms for
international mediation of claims is possible. The body of law regulating the interregional and international allocation of scarce water resources is huge, as is the
body of accumulated experience (Biswas, 1994c; Chitale, 1995). No international
framework yet exists for resolving water disputes, but work toward such a framework has begun. Mageed and White (1995) and Grover and Howarth (1991) review

existing institutional arrangements. Regional organizations, such as the Interstate Coordinating Commission for Water Resources (ICCW) in Central Asia and the Middle East Water Commission (MWC) currently provide the natural institutional setting for the development of such regimes; some experts (Grover and Biswas, 1993; Chitale, 1995) have called for the constitution of an integrated World Water Council. The potential of water scarcity to serve as a focus of conflict cannot be denied, but such conflict will likely be a measure of last resort.

Scarcity of other resources may also generate conflict. Myers (1989, 1993b) and Homer-Dixon *et al.* (1993; cf. Homer-Dixon, 1994) argue that competition for natural resources, worsened by demographic pressure, was to blame for the 1969 "soccer war" between El Salvador and Honduras. A series of case studies (Howard and Homer-Dixon, 1995; Kelly and Homer-Dixon, 1995; Percival and Homer-Dixon, 1995) has broadened application of the basic model of environmental scarcity and violent conflict. At the same time, the model according to which regional environmental degradation poses an international security problem has been subjected to a strong critique from within political science and international relations (Levy, 1995a, 1995b), and has been defended just as spiritedly (Homer-Dixon and Levy, 1995). The basic problem in this area is that many of the root causes that make countries prone to violent conflict also make them prone to degradation of the renewable natural resource base.

5.3.2   Impact of climate change

The research on agriculture, land, soil, and water surveyed earlier in the chapter gives reason for concern that renewable resource systems will become less reliable as a result of global climate change and that impoverished groups living in rural areas of LDCs are most at risk, particularly populations subsisting on crops grown on marginal soils or those living in semiarid regions. In addition, populations living in low-lying coastal zones will also be at risk of displacement due to rising sea levels, a wide-ranging impact reviewed in Chapter 1.

As the productivity of the renewable resource base declines, a bifurcation point is reached (Tiffen *et al.*, 1994; Mortimore, 1989). One branch corresponds to the environmental refugee scenario: worsening environmental deterioration, growing food scarcity, starvation and distress out-migration, dependence on international relief, etc. The other branch, which shares the spirit, if not the particulars, of the Boserup model of agricultural intensification (Boserup, 1965), is characterized by land-saving investment, adoption of improved agricultural practices, development of nonagricultural economic activities, diversification of income sources, etc. Droughts, which tend to extend over several years and recur regularly, give societies time to develop a range of adaptive responses. Floods, being acute events, are much more likely to provoke population dislocation. Displaced populations are likely to reestablish themselves once the waters recede. Thus, while floods are more likely to produce environmental refugees than are droughts, these persons' status as refugees will not necessarily be permanent or even long term. Although droughts cause less dislocation in the near term, they can cause more significant long-term social change because of their persistence.

The distinction between vicious and virtuous responses can be applied to the resolution of conflicts as well. Because environmental change forces choices which amount to the weighing and contesting of competing claims, conflict per se is essential to human adaptation and change (Stern *et al.*, 1992). Violent conflict is only one in a range of means by which competing claims to scarce resources can be settled. If social and political institutions for conflict resolution prove inadequate to the rising demands placed upon them, then regional environmental change associated with global warming might increase the frequency of skirmishes over impaired renewable resource systems. On the other hand, while access to renewable resource systems spanning national boundaries will be an important component of any country's foreign policy, all options will be explored before a country resorts to war. Suhrke (1993) concluded that environmental degradation was more likely to lead to exploitation of the poor and chronic low-level conflict than to acute conflict, let alone war.

Environmental stress is dealt with by institutions whose functioning is contingent on their perceived ethical legitimacy as well as on an agreed-upon and correct interpretation of the problem to be solved (Orstom, 1990). By necessity, populations cope in one way or another with environmental problems. At one end of the continuum is successful coping-in-place; at the other is population displacement under highly distressed conditions; somewhere in the middle lies a wide and complex variety of migration strategies. The means of coping in turn affect the nature of institutions. In the case of population displacement, this may amount to institutional fracture and atrophy; in the other cases, it may amount to institutional deepening and strengthening. The means of coping may also have a direct effect on the ecosystem under consideration and may alleviate or worsen sources of stress. The process is dynamic and path dependent (Lee, 1987; McNicoll, 1990, 1993); once started down the destructive path alluded to above, the process may be impossible to stop.

5.3.3 Population and environmental security

Links between population and the major environmental components of environmental security (soil, water, etc.) were discussed earlier in this chapter. To complement these discussions, it is appropriate to focus on the complex issue of migration. We examine first seasonal or temporary migration as a response to environmental degradation and then consider the larger-scale process of international migration.

Demographic Response to Ecological Stress

The typical demographic response to ecological stress (specifically, a run of bad harvests) was described in Section 2.6. In drought-prone regions, seasonal or temporary migration of some household members may already be an established means of coping with periodic food shortages; when disaster strikes, households resort to the coping strategy with which they are most familiar. This was the case during the 1980s drought in the Sahel, during which Hill (1990) found that the most prevalent coping strategy was to migrate. Findley (1994) found that in Mali, the drought did not cause an increase in migration per se, but rather led to a dramatic shortening of the periodicity of circular migration, leading to the conclusion that the appropriate policy response is to facilitate migration. Caldwell *et al.* (1986) looked at coping strategies employed by households in drought-prone southern India during an especially severe episode. By far the most common response was eating less (to some extent in all households; to the point of hunger in one-third of households), followed by reductions in discretionary spending (in just under one-fifth of all households), followed by securing loans, selling assets, and changing employment. When drought results in famine, institutional rigidities and failures are invariably to blame (Chaibva, 1996).

The migratory response to environmental stress depends in part on local conditions, such as the nature of land tenure, the status of women and children, the extent of poverty, and the enabling or impeding role of the state. The role of public policy is illustrated by Warrick's (1980) examination of the changing impact of recurrent droughts in the American Great Plains states. Great droughts in the 1890s and 1910s provoked large-scale out-migration; however, out-migration was much less pronounced in the 1930s and 1950s because public policies and programs to encourage adaptation-in-place were in force. Responses may, on the other hand, emerge spontaneously at the community level. In Rajasthan in western India, households that participated in voluntary common property management schemes were less likely to out-migrate than households that chose not to participate (Chopra and Gulati, 1998).

International Migration

Scenarios in which environmental stress leads either to virtuous out-migration or to environmental collapse and displacement of populations can easily be broadened to the national scale. International migration will be an important variable in any global environmental change scenario because it could provide a means of adjusting to changing climatic conditions. The response to global environmental change in a world with relatively free international movement of labor (and capital) is bound to be very different from the response in a world with closed borders.

While refugees tend to move within countries or across neighboring borders, economic migrants tend to move either from South to North, from East to West, or into regional immigration poles such as Abidjan and Singapore. Neoclassical economic theory describes how, in a two-region world, there are aggregate welfare gains in both regions when factors are free to move from one to the other (Simon, 1989; Layard *et al.*, 1992). Unimpeded, labor will flow from the low-wage, labor-abundant region (i.e., the region characterized by a low capital-to-labor ratio) to the high-wage, capital-abundant region, while capital will flow in the opposite direction. Economic scarcities are thus abated, with resulting increases in welfare in each region. The problem is that within each region there are winners and losers: the scarce factor of production loses and the abundant factor wins. The greater the internal economic rigidities, the greater the imbalance. When the analysis moves to the macroeconomic level, there is less clarity; for example, there is no consensus among American researchers on the fiscal impacts of immigration.

On the other hand, international migration cannot be reduced to simple economics, because it is driven not only by economic disparities, but also by more diffuse social and cultural factors. The diffusion of Western culture raises demand for a Western lifestyle; the rise of individualism requires mobility. Similarly, opposition to immigration is not entirely a matter of economic interest-group politics, it also reflects a fear that local cultural traditions will be eroded with the loss of homogeneity. All in all, the social consequences of immigration are far greater than the economic consequences.

There are several imbalances, and inequities, in the current international migration process. First, whereas intercontinental migration provided a major "escape hatch" for Europe during its period of rapid demographic expansion, Third World populations so far have benefited much less from migration (Emmer, 1993). Second, whereas policy reforms of the past 20 years have substantially reduced barriers to the flow of capital from industrialized countries to the South and to the East, the barriers to the flow of labor into industrialized countries have, if anything, increased (van de Kaa, 1993). Policymakers in LDCs, as well as in the formerly socialist transition economies, have put in place comprehensive policy regimes to facilitate capital inflows, whereas governments in industrialized countries have not adopted complementary policies in the area of immigration (Bade, 1993, for the case of Germany).

If the current research which assigns the most serious agricultural and health impacts to developing countries is correct, under conditions of global climate change there will be increased South–North migratory pressures. To the extent that LDCs are able to adjust smoothly to changing climate conditions, these pressures will be smaller, but perfect adjustment is an ideal that is unlikely to be attained. The North's adjustment to these pressures, as in many of the other areas explored, can be either constructive or destructive. In the latter case, immigration policy will remain essentially ad hoc and inequitable, in which case the inevitable confusion is likely to bring about an increase in the number of illegal immigrants. Newcomers will not be integrated into host societies, and broader domestic policies relating to problems such as structural unemployment will not be implemented, resulting in xenophobia and resentment. In the virtuous case, countries will elaborate comprehensive immigration policies that respect national priorities while preserving equity and transparency, immigrants will be integrated into their destination societies, and policies will be found to address the problems of vulnerable groups, such as unskilled workers.

5.3.4   Conclusion

The concept of "environmental refugees" has attracted wide interest. While its definition may be too broad, it reflects the general conclusion of research in the area: rising environmental pressures are a push factor encouraging out-migration (Lonergan, 1998), and if a virtuous out-migration response is stifled, then rapid environmental deterioration and resulting distress migration are possible. A similar observation applies to violent conflicts over natural resources: scarcity invariably gives rise to conflicts between stakeholders. If these are not resolved by institutions, including the market, then violent conflict becomes a distinct possibility. Stresses associated with global change will probably intensify the pressures that already drive internal, regional, and intercontinental migration, and policymakers and societies will be forced to come to terms, one way or another, with these rising pressures. Both virtuous and vicious adjustment paths are available, and it would be wrong to jump to the conclusion that global climate change presages a century of massive refugee movements and violent conflicts. However, lower fertility and slower population growth would contribute to relieving the proximate causes of distress migration (soil degradation, for example) and ease the institutional adjustments alluded to above.

5.4   Conclusion

The expected impacts of climate change could threaten agriculture, health, and environmental security, especially in LDCs. This does not mean that climate change is expected to make living conditions worse than they are today; rather, conditions are expected to be worse than they would have been in the absence of climate change. While living conditions in today's LDCs are bound to improve greatly over the time frame considered here, that improvement would be facilitated if there were no climate change.

Against the negative impacts of climate change must be set institutional and social responses through the market, government agencies, the legal system, household structure, etc. A case can be made that lower fertility at the household level and slower population growth at the regional and national levels would ease the challenges faced by these institutions. A qualification specific to the case of health is that lower fertility accentuates population aging and thus puts pressure on health resources. Another, general, qualification is that in none of the three areas considered are policies affecting fertility likely to be key strategies, since more direct means of improving resilience are available. Among these are better management of agricultural resource systems, more equitable distribution of available health resources, and elimination of rigidities that trap impoverished populations in environmentally unstable environments.

Notes

[1] Because they dominate world calorie supply, we concentrate in this section on the major cereal grains. However, among the ecosystems most profoundly affected by global environmental change may be the oceans, with significant implications for fish stocks (Mann and Drinkwater, 1994). Fish is a major source of protein for many populations. Climate change could potentially affect fisheries by altering the strength of winds and oceanic circulation and by changing climatic zones, which might affect reproductive patterns, migration patterns, and ecosystem relationships. Huge uncertainties are encountered in assessing climate change impacts on fisheries (Kennedy, 1990; Shuter and Post, 1990). Freshwater fisheries would also be affected by precipitation changes and are more sensitive to temperature changes than are ocean fisheries. In the near term, however, the outlook for world fisheries will be dominated by the impacts of the severe overfishing that has long been evident. Efforts by international policymakers to alleviate this overfishing have not succeeded to date.

[2] Authors (e.g., Myers, 1994a, p. 56) often cite statistics that illustrate a "slowing of gains" in the world food system. MacKellar and Vining (1987) have shown that such calculations are sensitive to the indicator chosen, the years compared, and the summary measure employed. Dyson (1991) offers a more systematic rebuttal of the case that world agricultural indicators paint a pessimistic picture.

[3] "Emergence of a highly developed international economy," writes Brown (1985, p. 7), "provides a way of transmitting scarcities from one country to another, a sort of domino theory of ecological stress and collapse." This rigid metaphor could not be further from economists' vision of the trading system as an elastic instrument permitting shortfalls to be balanced by surpluses.

[4] This is essentially the "Lewis model" of economic development with an environmental slant. In the Lewis model, rapid population growth is undesirable because it concentrates resources into agriculture, which is an inherently low-productivity sector. Here, the problem is that the intensive agriculture necessary to accommodate rapid population growth is environmentally unsound.

[5] Some researchers believe that extra income in poor households does not translate into higher nutrient intake but is spent on "junk" calories instead. Others argue that when initial nutritional status is included as a conditioning variable, the income elasticity of nutrient intake among the undernourished is quite high. Ye and Taylor (1995) found that, while total nutrient intake in northern China is inelastic with respect to income, consumers switch from grain sources to meat sources as income rises. This explains, in part, the evidently weak relationship between nutrient intake and income: as income rises, switching behavior also raises the cost of nutrients.

[6] In this section, we concentrate on global warming, despite the fact that related environmental changes, such as the depletion of stratospheric ozone, also have possibly significant impacts on agriculture. The impact on crop yields of low-level air pollution, such as increases in tropospheric ozone concentrations, is another serious problem that we do not discuss.

[7] In addition, there are several linkages that have been little studied, including effects on agricultural production caused by changes in the prevalence of pests and plant diseases (Cannon, 1998; Farrow, 1991; Sutherst, 1991).

[8] Some climatic events, such as flood or drought events defined in total precipitation terms, are nonnormal, although Katz and Brown (1992) speculate that similar increases in the frequency of extreme events would be observed.

[9] An example can be drawn from the Indian case described by Sinha and Swaminathan (1991). Droughts on a comparable scale affected the Indian subcontinent in 1965–1966 and 1987–1988. In the first, food grain production fell 18.8% from the previous year; in the second it fell only 8.7%. In the 1918–1919 drought, which was admittedly much more severe in hydrologic terms, food grain production declined by a third.

[10] Regarding irrigation practices, Smith (1994, p. 159) cites studies using a range of approaches that point toward increased runoff in the Aral Sea river basin. However, he characterizes as "unclear" the effects on actual water use in the region, which must be mediated by human interventions.

[11] By extension, the lower the scope for substitution, the more deleterious the impact. Thus, in a general equilibrium analysis, Winters *et al.* (1998) found that impacts of climate change on Africa are likely to be more severe than impacts on Asia or Latin America.

[12] Boserup was specifically concerned with African agriculture, in which case she argued that cultivation slowly shifts from forest fallow to annual multi-cropping due to increasing population pressure. Caldwell (1991) has argued that Boserup was concerned solely with exceedingly long-term agricultural evolution. This seems unlikely in view of Boserup's criticism of Evenson's (1988, 1993) even mildly Malthusian conclusions in his study of northern India; she argues that these were made inevitable because Evenson failed to endogenize government policy response in the form of programs to encourage intensification.

[13] For example, in the case of India several irrigation schemes were started in the second half of the 19th century while considerable land extensification continued until after the middle of the 20th century.

[14] Before World War II, when population growth in Machakos was slow, agricultural productivity was low and there was widespread land degradation in the district. But after the war, when the population of the district grew much more rapidly, this growth was accompanied by increased food production and reduced land degradation.

[15] Birdsall (1992) calculates from Allen and Barnes' (1985) cross-sectional regression results that the elasticity of deforestation with respect to population growth – absolute change in the average annual growth rate of forested area per unit absolute change in the average annual growth rate of population – is –0.5. Preston (1994) reports, based on World Bank data, a simple correlation coefficient of –0.32 between the national rates of population growth and deforestation.

[16] Off-farm costs, in the form of sedimentation of waterways, toxic runoff, etc., are, however, at least an order of magnitude higher than direct on-farm costs. This follows, in part, from the structure of relative prices: while food is cheap at present, people are willing to spend large sums of money for recreation.

[17] The United Nations Secretariat (1994, p. 31) provides a useful compendium of main-stream quantitative damage estimates: 70,000 square kilometers of farmland have been abandoned because their soils are no longer productive, another 200,000 square kilometers suffer from impaired productivity due to soil erosion. The productivity loss to land degradation is calculated to be equal to one-half of the actual increase in grain production each year.

[18] This conversion causes massive changes in riverine and surrounding ecosystems (Dudgeon, 1995, for the case of southern China), as well as frequent interregional and international political conflicts (Wescoat, 1991, for the case of the Indus River basin).

[19] Level of development also plays an important role here. In low-income countries, irrigation accounts for 90% of total water use; in middle-income countries, 70%; and in high-income countries, 40% (World Resources Institute data cited by World Bank, 1992, p. 100). The proportion devoted to domestic consumption is 4%, 13%, and 14%, respectively; the remainder is used for industrial purposes. Thus, while domestic use remains roughly fixed in proportional terms as countries move from middle- to high-income status, there is a decisive shift of water utilization away from agriculture and toward industry. A good part of this shift is due to improvements in the efficiency with which water diverted for irrigation is utilized.

[20] An order of magnitude of inefficiencies in domestic US water consumption is avail-able from comparing per capita daily water use of 105 gallons per day in homes served by public water systems with 78 gallons per day in homes served by private wells (Solley *et al.*, 1989).

[21] The demographer Jean Bourgeois-Pichat once invoked the image of "soft rock," in the form of mortality from infectious and parasitic disease (especially infant and child mortality) being chipped away, leaving the "granite" of degenerative disease. This model is now known to be misleading: infectious diseases cannot be eradicated, and there is nothing inevitable about degenerative disease (at least into advanced old age).

[22] See Dudgeon (1995, p. 37) for many references regarding river pollution in China, and Chourasia and Tellam (1992) for a study of the effect of irrigation on groundwater quality in central India.

[23] Climate change would also tend to increase various non-vector-borne infectious diseases, such as cholera, salmonellosis, other food- and water-related infections, and diseases caused by large parasites such as hookworm and roundworm. This increase would be most likely in tropical and subtropical regions, because of the effect of changes in water distribution and temperature upon microorganism proliferation.

[24] This finding is in line with the results of Martin and Lefebvre (1995), whose simulations indicated that while the total land area characterized by year-round endemicity would decline, the total land area characterized by seasonal endemicity would rise. The public health consequences of seasonal endemicity are more serious because of the larger nonimmune population.

[25] According to the 1951 United Nations Convention on the Status of Refugees and Stateless Persons, to be considered a refugee, a person must be outside his or her country of origin for reasons of "persecution" based on "race, religion, nationality, membership of a particular social group or political opinion."

[26] A related topic is the matter of the environmental impact of refugees on host environments (which may itself be a cause of conflict with the host society). Black (1994) found that empirical evidence was limited and causal processes were likely to be complex; Black and Sessay (1998) found no evidence that forced migrants resident in the Senegal River valley were more environmentally destructive than the local population.

Chapter 6

Population and Climate Change: Policy Implications

While recent international agreements have recognized relationships between population and global environmental issues, none has translated these linkages into specific recommendations. On the environment side, Agenda 21, signed at the Earth Summit in 1992 and intended as a blueprint for sustainable development, recommends only that nations take demographic factors into account in the policy-making process. On the population side, the Programme of Action agreed to at the International Conference on Population and Development in Cairo in 1994 also discusses population–environment links, but does little more than repeat the language of Agenda 21.

One logical forum for analysis of relationships between population and climate change is the Intergovernmental Panel on Climate Change (IPCC), which is charged with assessing the science of climate change and its potential impacts, as well as formulating response strategies. Yet it has paid little attention to population. For example, the IPCC's most recent reports on mitigation and adaptation options (Watson *et al.*, 1996; Bruce *et al.*, 1996) evaluate a wide array of strategies but do not consider policies to slow population growth. There are likely a number of reasons for this omission, not least the tension between North and South over the relative contribution of population and consumption to environmental problems (Bongaarts *et al.*, 1997).

In Chapters 4 and 5 we examined the role of population growth in generating greenhouse gas (GHG) emissions and its impact on the ability of societies to adapt to climate change. Chapter 4 concluded that under conditions of rapid demographic transition (lower fertility and lower mortality, resulting in slower population growth and an older population age structure) GHG emissions would be reduced relative to emissions in a baseline demographic scenario. While moderate in the short run, the

183

difference between the two emission paths was significant in the long run. The basic
conclusion that more rapid demographic transition translates into lower emissions
was found to be robust in a number of simple sensitivity tests. More direct means of
reducing emissions (such as improving energy efficiency) are available, and these
arguably have less pervasive social and economic effects than policies designed to
slow population growth. On the other hand, there is a case that the adjustment burden on these more direct policy interventions will be lighter in a world characterized
by slow population growth. Chapter 5, based in large part on vicious-circle models
described in Chapter 3, concluded that lower fertility would improve the ability of
developing countries to adapt to the expected impacts of climate change. Symmet-
rically, more direct policy interventions are available to strengthen institutions such
as markets, government agencies, the family, etc. However, the adjustment burden
on such institutions will probably be lighter in an environment of low fertility and
moderate population growth.

In this final chapter, we discuss these core findings in a policy context. First
we summarize current justifications for climate change and population policies.
We conclude the book by describing how climate change strengthens the case for
population policies.

6.1   The Basis for Climate Policy

As a policy issue, climate change is characterized by a difficult set of problems:
a long time horizon, which makes intergenerational issues central; global scope,
which implicates many nations with a range of social and economic systems, interests, and values; nonlinear and irreversible effects; scientific uncertainty; and
the necessity of addressing emissions of a number of gases from a wide range of
human activities. Policies responding to climate change are based on a number of
principles, varying from economic efficiency to equity concerns and rights-based
arguments. The IPCC, for example, concluded that climate change policies can
be justified on three grounds. First, some policies that would reduce emissions
or enhance adaptation are costless or beneficial even when climate change is not
taken into account. Such policies are known as "no-regrets" strategies, since no
one would regret their implementation even if climate change impacts turn out to
be negligible. Policies that are clearly beneficial for reasons unrelated to climate
change are often described as "win–win" policies, since they have both climate-
and non-climate-related benefits. Second, there is a reasonable scientific expectation that climate change will inflict significant costs on society. Third, individuals
are willing to pay a reasonable price for precaution, even when the underlying risks
are uncertain. Another possible policy justification is that climate change may make
global inequities worse than they otherwise would be.

6.1.1   No-regrets strategies

The IPCC has singled out no-regrets policies (defined above) as high-priority mitigation responses (Bruce *et al*., 1996). There is disagreement over the extent to which such opportunities exist. There are two main schools of thought (Hourcade, 1996). The first reflects the engineering point of view embodied in "bottom-up" model-based technology assessments. Such studies find that current energy services are less efficient than would be possible with the best available technology, and that adoption of more efficient equipment could lead to substantial emission reductions at very low or even negative costs. The most optimistic proponents of this school of thought (Lovins and Lovins, 1997) assert that opportunities for improving energy efficiency are so bountiful, and so profitable, that the climate debate is largely irrelevant; energy-saving measures should be undertaken solely for their return on investment. Firms do not capitalize on these opportunities because they face a number of obstacles, including perverse incentives, lack or poor quality of information, poorly designed regulatory regimes, and distorted price signals.

The second school of thought sees little scope for no-regrets measures, reasoning that businesses would already have taken advantage of such opportunities if they existed, or that the cost of removing obstacles to them would outweigh the benefit that would be realized. This view is embedded in "top-down" models based on aggregate behavior and macroeconomic indices, which therefore produce higher estimates for the costs of reducing GHG emissions.

An IPCC review of the existing literature concluded that significant no-regrets options exist in most countries (Bruce *et al*., 1996). Energy subsidies, for example, introduce price distortions that can lead to inefficient energy use. Poorly defined property rights can lead to deforestation. These kinds of market and institutional failures provide opportunities to implement corrective policies that would realize GHG emission reductions at a net economic benefit even in the absence of damages from climate change.

6.1.2   Expected costs of climate change

Policy measures designed to reduce GHG emissions are justified on economic efficiency grounds if their costs are less than the resulting benefits. Because the impacts of climate change are expected to be, on balance, negative, some level of investment in mitigation policies beyond implementation of no-regrets policies is justified. Cost–benefit analysis provides a tool for determining what the level of investment in GHG emissions should be; in principle, investments should be made up to the point at which the marginal cost of mitigation just equals the marginal benefit. Reducing emissions further would cost more than the value of the benefits society would enjoy as a result of the reductions. Reducing emissions less would also be inefficient, since by spending an extra dollar on reductions, more than a dollar's worth of benefits could be realized.

Because the damages caused by GHG emissions are not reflected by the price system, some form of intervention is required to achieve an economically efficient level of emission abatement. One such intervention is to set a tax on the carbon content of fuels. Other interventions, such as the cap on total emissions combined with a system of tradable emission permits called for in the Kyoto Protocol, can achieve the same result.

Cost–benefit analyses of climate change must confront a number of formidable difficulties. Estimates of costs and benefits are subject to great uncertainty. For example, estimates of benefits must cope not only with uncertainties in climate change impacts, but also with uncertainties in putting a price tag on them (Pearce *et al.*, 1996). All costs and benefits must be expressed in economic terms, yet many benefits are difficult or impossible to quantify. Valuing human health has proved particularly difficult and controversial, as has valuing nonmarket goods such as biodiversity and environmental amenities. Values for such goods include direct and indirect use value (e.g., visits to parks, or the medicinal value of plant species) as well as non-use value such as option value (e.g., preserving species that may have use value in the future) and existence value (e.g., the value of knowing that certain species still exist). Estimating non-use value presents special difficulties and theoretical challenges (Attfield, 1998; Green and Tunstall, 1991) and is generally based on contingent valuation methods such as surveys of the public's willingness to pay for goods or willingness to accept compensation for their loss. As a result, most studies have focused disproportionately on the more easily quantified direct impacts of climate change, such as impacts on agriculture and the consequences of sea level rise (Fankhauser, 1994).

Nonetheless, cost–benefit studies can provide a useful benchmark for decisions on emission reductions in the short term (Portney, 1998). While models containing a range of cost and benefit assumptions produce widely varying estimates of optimal emission reductions in the long term, estimates of current optimal reduction rates do not rise much beyond 10% relative to a business-as-usual scenario (Weyant, 1996).

6.1.3 Risk aversion and the precautionary principle

One justification for climate change mitigation in excess of these modest sums is risk aversion. Individuals are said to be risk averse when they are willing to pay in excess of an actuarially fair premium to purchase insurance against risk. For example, if unabated GHG emissions were expected to lead to a present value loss of 2% of gross domestic product (GDP) with 50% probability, then based on a risk neutral expected-value approach, mitigation costs of 1% of GDP (2% $\times$ 50%) would be justified. However, if societies are averse to the risk of damages, additional investments in mitigation would be justified. Risk aversion is widely observed in the real world. Analyses of sequential decision-making strategies under conditions of risk aversion demonstrate that optimal actions hedge against the risk of future damages by taking more action now than would be the case if agents were risk neutral (Hammitt *et al.*, 1992; Manne and Richels, 1992).

Uncertainty is rather more complicated than risk. Wynne (1992) distinguishes four types of uncertainty: indeterminacy, in which outcomes are in principle unknowable; ignorance, which describes outcomes outside the range of those currently defined as possible; classic uncertainty, in which outcomes have been identified as possible but their probability is unknown; and risk, in which the probability of outcomes is known. The simple numerical example above deals only with risk, the most tractable type of uncertainty. Climate change more commonly involves the other three, more difficult types of uncertainty. For example, local climate projections are indeterminate. "Ignorance" describes possible consequences of climate change that have not yet been thought of. Although by definition examples cannot be provided, the ozone depletion issue presents a parallel instance in the formation of the Antarctic ozone hole, which was entirely unexpected. Many, if not most, impacts of climate change fall under the category of classic uncertainty as defined above, high-consequence events of this sort include the possible "surprises" discussed in Chapter 1, such as shutdown of the North Atlantic ocean circulation – its possible occurrence has been identified, but its probability is unknown. Fewer aspects of climate change projection fall into the the risk category: projections of GHG concentrations over intermediate time scales, given assumed emission scenarios, might be one such aspect; the better understood parameters of the climate system might be others.

When faced with such an array of uncertainties, a value choice must be made in deciding how to act. The past decade has seen growing application of the precautionary principle to such situations. This principle states that society should err on the side of caution; that is, rather than waiting for uncertainty to be resolved, society should take action to ensure that possible environmental impacts are minimized. The precautionary principle shifts the burden of proof away from those seeking to protect the environment and toward those whose actions might harm it. The Framework Convention on Climate Change (FCCC) explicitly states that the precautionary principle is to be followed: "The Parties should take precautionary measures to anticipate, prevent, or minimize the causes of climate change and mitigate its adverse effects. Where there are threats of serious or irreversible damage, lack of full scientific certainty should not be used as a reason for postponing such measures … ."

The precautionary principle has become a widely invoked principle in international environmental resolutions, appearing not only in the FCCC but also in the Convention on Biological Diversity, the Rio Declaration on Environment and Development, and in amendments to the Montreal Protocol on Substances that Deplete the Ozone Layer, as well as a number of regional treaties (Cameron, 1994). However, it has been criticized as an insufficient basis for policy because it provides no guidance on crucial questions of how much (if any) precautionary action should be taken (Bodansky, 1991; Dovers and Handmer, 1995). The principle can be used, for example, to justify a complete shifting of the burden of proof so that no activity may be undertaken until it is proved safe to the environment. On the other hand, it can also be judged consistent with a mere relaxation of standards of definitive proof of environmental harm before regulatory action can be taken. Similarly, the principle says nothing about how to make decisions on the magnitude of costs that are justified in avoiding possible impacts. In the case of climate change, the FCCC does little to help matters by calling for application of the precautionary principle when "serious or irreversible damage" is threatened. What constitutes "serious" or "irreversible" damage and how much (and what type of) uncertainty is tolerable? Finally, the precautionary principle does not provide guidance on what kind of action to take. Precautionary policies in the climate change field include not only reductions in GHG emissions, but also research on climate change and impacts, development of new energy supply and conservation technologies to reduce abatement costs, and policies to improve the ability of societies to adapt to climate change.

Despite its shortcomings, the precautionary principle serves as a useful general principle having legal effect in a number of instances, including the FCCC (Cameron, 1994). Ultimately, no single approach to climate change policy will be universally applied. Risk can most effectively be coped with by constructing a portfolio of responses that includes both mitigation and adaptation strategies. Investing in an array of measures is likely to lead to a greater return than focusing on a single strategy, and choice will differ from country to country depending on national circumstances. The IPCC has presented a list of possibilities which run from fuel switching and removal of subsidies for fossil fuel use to research, international cooperation, and institutional development.

6.1.4   Equity

Changes in welfare due to the impacts of climate change will not be distributed equally; there will be differences across geographic regions, economic sectors, and generations. Addressing questions of equity between current and future generations or between better- and worse-off regions of the world requires an appeal, whether implicit or explicit, to ethical criteria and can provide additional justification for climate change policies.

For example, the balance between costs and benefits will shift over time, since costs of reducing GHG emissions must be borne in the near term, whereas the benefits produced, in terms of avoided damages, are primarily long term in nature. Deciding what costs the current generation should bear in order to produce benefits for future generations involves a value judgment, and this judgment can be based on a number of different principles. Cost–benefit analyses generally seek to maximize total well-being over time. To compare costs and benefits at different times, a discount rate is used; the higher the discount rate, the less the well-being of future generations is valued relative to that of the present generation. The choice of discount rate has a strong effect on the outcome of cost–benefit analyses and is a matter of controversy (Portney and Weyant, 1999; see *Box 3.1*). An analysis using a low discount rate will call for greater present emission reductions than will an analysis using a high discount rate.

Different conceptions of equity across space, as well as time, can also affect the decision of how to address the threat of climate change. Because less developed countries (LDCs) depend more heavily on agriculture, and because they have generally less developed institutional infrastructures for responding to climate change, they are likely to be more vulnerable to damages than are more developed countries (MDCs). Climate change mitigation should benefit LDC populations more than MDC populations, and thus may be considered a transfer justified on equity grounds. The fact that the global population distribution is tilting toward the South strengthens this rationale.

International equity is central to the issue of who should bear the costs of mitigation. The FCCC employs a form of the "polluter pays" principle in addressing this issue in Article 3.1: "The Parties should protect the climate system for the benefit of present and future generations of humankind, on the basis of equity and in accordance with their common but differentiated responsibilities and respective capabilities. Accordingly, the developed country Parties should take the lead in combating climate change and the adverse effects thereof." Thus, the basic feature of the Kyoto Protocol is that industrialized countries, but not developing countries, are called upon to reduce emissions. One of the main issues in international climate politics and diplomacy is how much (and when) LDCs should also be called upon to contribute to the stabilization of GHG concentrations.

6.1.5 Summary

Justifications for climate change policies can be based on four principal lines of reasoning. First, the existence of no-regrets options provides a basis for GHG emission reductions in order to realize benefits unrelated to climate change. Second, the expectation that, on balance, climate change impacts will be negative justifies on economic efficiency grounds policies that would carry some net cost in the absence of climate change. Third, the precautionary principle asserts that policies should err on the side of caution when it comes to potentially serious or irreversible damages. Fourth, inter- and intragenerational equity considerations provide an ethical basis for climate change policy.

6.2   The Basis for Population Policies

All three components of population change – fertility, mortality, and migration – are subject to public policy. However, in view of the critical role of fertility identified in Chapters 2 and 3, we concentrate here on policies and programs related to fertility. These include family planning (FP) policies; maternal and child health (MCH) initiatives; information, education, and communication (IEC) programs; and a range of general policies designed to improve the status of women, including those to encourage the education of girls.

Two broad goals can be identified: narrowing the gap between desired and actual fertility and fostering changes in desired fertility. Some policies and programs address only one goal. For example, FP programs that supply contraceptives enable couples to achieve desired fertility, whereas IEC programs may be designed to influence desired family size. Other policies and programs simultaneously reduce the gap between actual and desired fertility and lead to changes in values. For example, the research cited in Chapter 2 shows that women with higher levels of education have lower desired fertility and are more likely to use contraception than those with lower levels of education. Policies such as those that improve the status of women or educate teenagers about sexuality and family life affect fertility only indirectly but, *ipso facto*, should be considered relevant to population policy.

Virtually all governments agree that measures that help couples to attain desired fertility, and that help to foster an equitable decision-making process within the family, are justified – hence, the universal acceptance of voluntary family planning programs. However, virtually all governments are reluctant to interfere overtly in families' fertility decisions; thus, policies that encourage changes in fertility goals must clear a higher hurdle. One justification might be that the policy in question – say, education of girls – is so virtuous in its own right that it should be implemented whatever its impact on fertility norms. The impact on desired fertility might, nonetheless, be judged to strengthen an already strong case in favor of the policy. Another justification might be that there is an externality to fertility decisions – that is, a cost or benefit to society as a whole or to future generations that is not taken into account by the decision-making couple. We discussed two such externalities, the GHG externality to high fertility and the aging externality to lower fertility, in Chapter 3.

The consensus on enabling couples to have the number of children they desire is reflected in the Cairo Programme of Action, the statement that emerged from the 1994 International Conference on Population and Development. The Programme justifies population policies primarily on the basis of empowerment of women and the securing of reproductive health rights. Impacts on fertility, and resulting impacts on demographic variables such as population size, rate of growth, and age structure, are considered in the Programme to be only a secondary motivation for population policy (Sen, 1995). Links between fertility and demographic impacts on economic development and the environment receive little attention.

It is worthwhile to put the Cairo Programme into context. In the postwar period there have been five rationales for fertility-related policies: those based on geopolitics, macroeconomic consequences of population growth, externalities to childbearing, ecological concerns with scale, and individual welfare. The relative importance of these rationales has changed through time and varied by country.

6.2.1 The geopolitical rationale

The history of international population policy is inextricably linked with the history of the international family planning movement, which emerged in the 1920s as a complex amalgam of eugenicists, public health advocates, and social reformers (Adams, 1990; Hodgson, 1991). Scientific advances in genetics and large-scale human rights abuses in Germany under national socialism thoroughly discredited both eugenics and policies derived from it. As a result, the international population movement was in disarray after World War II (Kühl, 1997).

In reconstructing itself, the early postwar movement relied heavily on two justifications for public support of voluntary family planning. One justification was the health and well-being of women, children, and young families. The second justification, not stated openly but never far in the background, was geopolitical in nature. During the 1950s, new census data revealed staggering rates of demographic increase in India and other LDCs. With the Cold War as background, policymakers in the West feared that exploding Third World populations would be fertile breeding grounds for political instability. Early US government international assistance for family planning was explicitly justified in these political terms (Donaldson, 1990).

Concerns over national economic and political standing have also sometimes served as the basis for pronatalist policies, such as generous parental leave, child allowances, public provision of day care, etc. (Teitelbaum and Winter, 1985). In the 1960s and early 1970s, for example, a number of East European countries, alarmed by the prospect of population decline, restricted abortions and put in place financial incentives for childbearing. With many MDCs facing labor force shrinkage and even possible overall population decline, pronatalism may be on the rise in coming decades.

6.2.2   The macroeconomic rationale

At the Bucharest population conference in 1974 (see *Box 6.2*), MDCs argued that higher fertility in LDCs was an impediment to macroeconomic growth. By the time of the next international population conference, held in 1984 in Mexico City, virtually all countries with the exception of the United States had adopted this view.

In Chapter 3 we discussed the long-standing argument that there is an inverse relationship between the rate of population growth and the rate of economic growth. Empirical research does not provide compelling support for this proposition. Nonetheless, the perceived negative economic consequences of high fertility have been one rationale for both national population policies and international population assistance.

The macroeconomic consequences of population aging, while uncertain, are of increasing concern to MDCs, which must cope with overcommitted pension and health systems. While no country has explicitly justified pronatalist policies in terms of population aging, this rationale may be at the back of policymakers' minds. Impacts of population aging on LDCs are an emerging policy concern.

6.2.3   The welfare–economic rationale: Externalities to childbearing

As discussed in Chapter 3, externalities are defined as consequences of fertility falling on society as a whole (or on future generations) that are not taken into account by parents when they decide how many children to have. In the case of a negative net externality (net social benefits of a birth being less than net private benefits), the magnitude of the externality represents the per-child tax that a government should theoretically levy on parents to maximize social welfare. In the case of a positive net externality (net social benefits of a birth being greater than net private benefits), the magnitude of the externality represents the payment that a government should theoretically make to parents. More generally, a range of incentives and disincentives to childbearing can be employed to cope with the externality problem. Generous West European family policies, while publicly framed in terms of social welfare, not demography, may be interpreted in part as a means of coping with the positive net externality to childbearing in societies characterized by sub-replacement fertility. Negative externalities to childbearing in LDCs would enhance the return to population policies, including those that act indirectly to reduce fertility, such as improved maternal and child health.

More controversial are direct antinatalist incentives, which have been employed in a small number of LDCs (Chomitz and Birdsall, 1991). Taxes on the number of children are likely to hurt the poor more than the wealthy, and family-size penalties can unfairly penalize children of higher birth order. Incentives for sterilization, which have been utilized in Bangladesh, India, and Sri Lanka, raise the issue of entrapment; poor women may accept payment for sterilization but later regret the decision. More generally, incentives may be viewed as unethical because they inject financial considerations into an area that involves fundamental human rights.

6.2.4   The ecological rationale: Scale and ecological concerns

The scale of human activity, as the I=PAT identity illustrates, depends on both population size and per capita resource use. McNicoll (1995), not to mention many ecological economists, points to scale as a potential flaw in neoclassical approaches to issues involving demographic and economic futures. Because the neoclassical model assumes infinite substitutability, it precludes the possibility of absolute resource scarcity. Demeny (1986) makes a similar point, arguing that the inability to deal with long-term population change causes the neoclassical approach to foreordain its conclusions.

The population component of scale gives rise to the concept of carrying capacity, as reviewed in Chapter 3. Carrying capacity, in turn, may point to the adoption of goals and targets for total population size. China, for example, has adopted targets for total population size as one basis for its population policies, which involve birth quotas, disincentives for large families, and incentives for small families. However, some have accused the Chinese approach (more specifically, the one-child policy) of being in violation of human rights.

There is no agreed-on formula for determining a sustainable population size. From the ecological point of view, the range of global carrying capacity estimates has widened, rather than narrowed, over time (Cohen, 1995). A small economic literature on optimal population size has made welfare comparisons between populations of different sizes but has not come to any clear conclusions (Dasgupta, 1986, 1998).[1] On the positive side, rapid population growth can give rise to economies of scale and what, in Chapter 3, we called Boserup–Simon effects in the form of accelerated technological progress. The empirical evidence in this area has never been strong enough, however, and the time horizon of policymakers is too short for such effects ever to have served as a basis for policy.

6.2.5   The individual welfare rationale

Over the past decade the view that policies affecting population should above all stress the welfare of individuals, as opposed to demographic trends and their macro-consequences, has gained primacy. In this view, ensuring access to health care and education, fostering women's empowerment, and guaranteeing reproductive rights are ends in themselves, not means to achieving particular demographic goals (Bok, 1994). Overemphasis on "numbers," according to this school of thought, has led to coercive practices and insufficient attention to contraceptive safety, with consequences falling largely on women. Voluntary family planning programs, for example, are now seen as a means of guaranteeing the rights of individuals and couples to have the number of children they desire, not as a means of reducing fertility. They are viewed as just one part of what should be a broad provision of reproductive health services that also includes pre- and postnatal health care, treatment of infertility and other reproductive health problems including sexually transmitted diseases, and education and counseling on all aspects of pregnancy, childbirth, infant and women's health, and parenting.

Population policies are also justified in this view if their aim is to improve child well-being. In some settings, as discussed in Chapter 3, high fertility can set off a self-reinforcing series of responses at the household level that impairs the health and education of children. If parents are not fully informed about the costs of high fertility to children (a form of intergenerational externality), or if the parents' need for the economic contributions made by children is stronger than their altruism toward their offspring, then intervention may be justified to alleviate the conditions leading to high fertility (Birdsall and Griffin, 1993).

The individual welfare rationale heavily influenced the Cairo Programme of Action (McIntosh and Finkle, 1995). In it, voluntary choice in matters of childbearing was enshrined as a basic right and a wide range of initiatives addressing reproductive health was called for. Gender issues, such as improving the status of women through education and economic opportunities and securing reproductive rights, were given prominence.

Some have lamented the lack of urgency within the Cairo document concerning rapid population growth in LDCs and its consequences (Westoff, 1995). The dominance of women's issues came at the expense of the traditional focus on the demographic rationale for population policies, and it has been argued that this shift risks loss of support for international population programs (Harvey, 1996). Environmentalists pointed out the lack of attention to the environment in the Programme of Action (*Earth Negotiations Bulletin*, 1994); a page or two of text relating to the environment contains little of substance beyond generic calls for integration of environmental, economic, and demographic concerns.

Yet the Programme of Action does set a number of quantifiable goals for the target year 2015: the elimination of unmet need for contraception; reductions in infant, child, and maternal mortality; increased life expectancy; and universal completion of primary education. In keeping with its focus on women's welfare as opposed to demographic consequences, there is no discussion of what impact the attainment of the targets might have on global population size, except to say that it would result in a global population below the UN medium projection. The Programme of Action also calls for increased spending to achieve these goals. It is estimated that a total of US$17 billion annually by 2000, rising to over $20 billion by 2015, will be required to fund the basic reproductive health programs outlined in the Cairo document. One-third of that amount is to come from international donors and the rest from developing countries themselves. Based on current trends, however, only a fourth of this amount is likely to be provided (Ashford and Noble, 1996).

Achieving the aims of the Cairo document would likely have the effect of reducing fertility and slowing population growth, so common ground probably exists between those most concerned with individual rights and those most concerned with consequences of population growth. The focus on reproductive health, which encompasses family planning, provided one point of agreement between the two camps (Chen *et al.*, 1995). But attempts at compromise will have to ease suspicions that those who have historically emphasized demographic targets are simply clothing the same objectives in new language (Sen, 1994b). Women's groups may be wary of, for example, the possibility that reproductive health is still seen as only those aspects of health that impinge on fertility. Similarly, little progress will have been made if unmet need for family planning is addressed by providing more of the kinds of services currently in place, rather than by expanding their focus from fertility reduction to a broad range of reproductive health services and involving women in their design and evaluation.

6.2.6   Summary

Population policies include a wide range of policies that directly or indirectly affect fertility, mortality, or migration. In the postwar period, justifications for programs in these areas have drawn on one or more of five basic rationales:

- A geopolitical rationale that views rapid population growth as a threat to national or global security and in some cases views slow population growth as damaging to national economic and political standing.
- A macroeconomic rationale that views population growth, and in some cases aging, as a barrier to economic growth and improvement in living standards.
- A welfare–economic rationale that identifies externalities that drive a wedge between the marginal social costs and benefits of private fertility decisions.
- An ecological rationale that views population size as a possible threat to the carrying capacity of natural systems.

- An individual welfare rationale that argues that population policies should be undertaken, not because of their demographic consequences, but primarily because they improve individual well-being. The Cairo Programme of Action, emblematic of this view, defines population policies broadly to include health, education, empowerment of women, and reproductive rights.

6.3   Climate Change and Population Policies

Climate change raises a range of issues that resonate with the justifications for population policy given above. GHG emissions associated with parents' fertility decisions represent a classic intergenerational externality. Surprise events and the high level of uncertainty revealed by the research summarized in Chapter 1 enhance scale concerns about human population size. Even the geopolitical motive for population policy, dated as it sounds, is not far removed from the debate over climate change and environmental security. In the remainder of this chapter we integrate the discussion in Sections 6.1 and 6.2, as well as the review and analysis in previous chapters, to describe how climate change strengthens the case for population policies.

Although population policies can affect a number of aspects of fertility, mortality, or migration, we focus here mainly on whether climate change strengthens the case for fertility-related policies. Policies impinging on mortality and migration are not unimportant, and the case for them is bolstered by taking climate change into consideration. For example, climate change would certainly strengthen the case for sound migration policies to the degree that projected stresses on natural resources can be expected to increase migratory pressures. International migration policies that enhance, rather than block, adaptation to climate change have an important contribution to make. They can also serve an equity goal by making available to LDCs the same migration safety valve that was available to today's MDCs at an earlier stage of their development. Similarly, mortality-related policies that improve health are likely to make societies more resilient to the impacts of climate change. However, Chapters 4 and 5 focused mainly on how mitigating or adapting to climate change is affected by the rate of population growth and age structure, which are mainly (although not exclusively) a function of changes in fertility. Mortality- and migration-related policies are less directly related to climate change.

Policies that lead to lower fertility, whether in LDCs or MDCs, are likely to lead to less climate change. In many LDCs that are currently experiencing rapid population growth combined with low standards of living, such policies are also likely to improve the resilience of societies and institutions to climate change. However, a number of other policy choices are available to address climate change, and they may be more effective. Compared with policies that have a direct impact on GHG emissions (such as development of technologies that are less carbon intensive), population policies are a blunt instrument, having broad impacts throughout the social and economic system. Some of these impacts, such as improvements in maternal and child health, are positive; others, such as possible impacts on technical change, are not well understood and could be either negative or positive. In a world of dense linkages, high uncertainty, and scarce financial resources, the fact that policies to promote slower population growth would alleviate the climate problem does not necessarily mean that they are the best approach. The unavoidable link between slower population growth and population aging deserves special attention. Lower fertility, especially in MDCs, could exacerbate problems associated with the rapidly rising elderly dependency ratio. But the same rules apply: the fact that maintaining current fertility levels or encouraging higher ones might alleviate problems of population aging does not mean that such policies are the best approach, especially in a world where population growth is contributing to climate change. More direct policies related to the reform of pension and health systems are also available.

Policy documents that have considered evidence on population growth and GHG emissions have arrived at different recommendations. For example, the United Nations Population Fund (UNFPA, 1991) concluded that population growth has been a principal contributor to increases in GHG emissions over the past several decades and will play a central role in future emissions growth.[2] Along with similar conclusions the report drew regarding the role of population growth in other environmental issues, this result led it to call for renewed efforts to reduce fertility.

In contrast, Rahman *et al.* (1993) sought to counter what they called "a potent myth" that assigns undue importance to population as a factor in climate change, concluding that reducing per capita emissions in industrialized countries would be a more effective and more ethically sound response. They emphasized that population momentum limits the impact that slowing population growth could have on GHG emissions over the next several decades; focusing on the population variable is therefore a distraction from the more pressing issue of reducing consumption in MDCs, which would create "environmental space" for developing countries.

The 1991 US National Academy of Sciences report (NAS, 1991) on the policy implications of climate change briefly examined the role of global population growth in producing emissions and in affecting the resilience of societies to climate stresses. It considered whether supporting economic development programs might be "self-defeating" from the point of view of climate change, since it would result in higher emissions. A scenario analysis concluded that more rapid economic growth (relative to a baseline scenario) in LDCs would not lead to higher emissions because it would drive a corresponding reduction in fertility. The report therefore called for support for broad economic development as well as increased funding for voluntary family planning services in LDCs, along with health and education programs.

Finally, as discussed at the beginning of the chapter, the most comprehensive, and certainly the most influential, scientific inputs to the climate change policy process are the reports of the IPCC, and they have never explicitly considered links between population policy and climate.

The paucity of scientific analysis of population–climate linkages has contributed to the lack of attention received by the issue at the policy level. As reviewed in *Boxes 6.1* and *6.2*, the record of high-profile international conferences on population or environment over the past several decades contains only generic statements on population–environment links, not recommendations based on analyses of specific issues. Even basic demographic factors have not always been incorporated into climate policy. Meyerson (1998), in analyzing the Kyoto Protocol, points out that accounting for varying population trends among MDCs (a factor overlooked at Kyoto) translates similar absolute emission caps into widely varying per capita limitations. For example, the population of the European Union is expected to increase by less than 3% between 1990 and 2010, while the US population is expected to increase by 18%, so that although the two regions agreed to nearly identical reductions in absolute emissions, US reductions in per capita emissions must be nearly twice as large (21% versus 11%).

6.3.1 Population policy as a no-regrets strategy

If the benefits of policies that slow population growth outweigh the costs even in the absence of climate change, then they can be considered no-regrets policies from the climate change point of view. As discussed at the beginning of the chapter, such measures have been singled out by the IPCC as priorities for climate change policy. Two specific fertility-related policies stand out as clear examples of no-regrets policies: reducing the unmet need for family planning services and investing in female primary education.

Reducing the Unmet Need for Family Planning

The existence of unmet need for contraception provides the most widely accepted basis for provision of voluntary family planning services. Few would argue with the goal of enabling couples to have the number of children they desire, and this goal is explicitly written into the Cairo Programme of Action. Policies to promote voluntary family planning would be desirable even in the absence of climate change, or even if climate change were to have no negative consequences. To the extent that such negative impacts exist, however, the return to family planning policies is enhanced. Unmet need for family planning and other reproductive health services

Box 6.1. Population in international environmental policy documents.

The United Nations Conference on the Human Environment held in Stockholm in 1972 laid the foundation for UN activities on environmental issues for the next two decades. The conference declaration (United Nations, 1973) helped elevate environmental concerns in the international arena, calling for their integration into economic and social planning. Yet few national reports submitted to the conference discussed population–environment issues (with the exception of concern over the consequences of rapid urbanization), and the declaration itself took no position on the relationship between population growth and the environment.

During the 1980s, more attention began to be focused on demographic factors. The report of the World Commission on Environment and Development, "Our Common Future" (WCED, 1987), is well known for establishing environment as a component of development by producing an often-quoted definition of sustainable development as "development that meets the needs of the present without compromising the ability of future generations to meet their own needs." The report also drew attention to the role of rapid population growth as a factor inextricably linked to poverty and environmental degradation. Although it made no specific population-related recommendations – and in fact its authors remained divided over the significance of population growth as a cause of environmental stress – the report boosted the prominence of population–environment linkages.

In the 1990s, growing attention was given to population–environment concerns, although still without analysis of specific linkages. The United Nations Conference on Environment and Development held in Rio de Janeiro in 1992 adopted three relevant agreements: the Rio Declaration, Agenda 21, and the Framework Convention on Climate Change (FCCC). Yet while 70% of the national reports to the conference cited demographic factors (particularly urbanization) as one of the causes of environmental concern (United Nations, 1997), none of the agreements made concrete recommendations regarding population-related policies. Population was mentioned in the Rio Declaration only in general terms, recommending that "States should reduce and eliminate unsustainable patterns of production and consumption and promote appropriate demographic policies." Agenda 21, a lengthy blueprint for sustainable development, was more specific, devoting a chapter to demographics and sustainability. It recommended integrating action on demographic factors with resource management and development issues, and incorporating measures to bring about demographic transition.

Despite these general proclamations, population did not appear at all in the FCCC.

Box 6.2. Environment in international population policy documents.

The World Population Conference held in Bucharest in 1974 brought population issues to the forefront of international concern. At the time, population's relationship to economic growth was the overriding concern. Nevertheless, considerable debate concerning the impact of population on the environment took place as well, with concern centered on depletion of nonrenewable resources and food scarcity (United Nations, 1997). The Plan of Action that was eventually adopted, however, made no prominent reference to the issue, noting only that national population goals and policies should consider natural resources and food supply along with more established considerations of economic and social factors.

At the International Conference on Population held in Mexico City in 1984, delegates met to discuss progress on and obstacles to implementation of the Bucharest Plan of Action. While links between population and economic development continued to dominate the agenda, environmental issues were given more prominence in the conference declaration than in the Plan of Action, with concern shifting somewhat from local to global impacts. The declaration recommended that, in countries where imbalance existed between population growth and the environment, governments should "adopt and implement specific policies, including population policies, that will contribute to redressing such imbalance."

The International Conference on Population and Development in Cairo in 1994 produced a new Programme of Action. As discussed earlier in the chapter, its emphasis was on individual rights, particularly the rights of women. Environmental considerations were perhaps more prominent than in earlier population documents, but essentially relied on language appearing in Agenda 21, which called for integration of population policies into sustainable development programs in broad terms.

---

is highest in LDCs, where the ratio of available resources to population in need is lowest. However, there is also significant unmet need in MDCs (including the formerly socialist countries in which health care systems have deteriorated), where the high level of GHG emissions per capita enhances the no-regrets aspect of family planning policy. In the case of LDCs, slower population growth is also likely to increase the ability of societies to adapt to climate change.

Climate change, in brief, further strengthens the already strong case for family planning policies. The potential impact of family planning policies on GHG emissions is significant: demographers estimate that eliminating unmet need for contraception in LDCs could reduce LDC population in 2100 by approximately 2 billion people (Bongaarts, 1994b). However, unlike the case of many no-regrets policies in the energy domain, the GHG emission reductions that arise from slower population growth will be obtained only in the long run. Nonetheless, it can be argued that climate change is an inherently long-term problem, in which case responses that have long-term returns should not be excluded.

Investing in Female Education

As discussed in Chapter 2, the inverse correlation between fertility and female education far outweighs correlations between fertility and other social variables such as income or labor-force participation. While correlations overstate causal relationships, in this case they provide a reasonably robust guide to policy. The strongest intermediate link in the causal chain is that the likelihood of contraceptive use in marriage rises with educational attainment. There may be increasing returns in terms of lower fertility to each additional year of schooling, but even at the primary school level the effect is strong. In addition to this direct link, female education affects all of the main factors that together lead to fertility decline: socioeconomic development, empowerment of women, and the availability and success of family planning programs.

Female education is both a desirable goal in its own right and, because it is strongly linked to lower fertility, an appealing indirect population policy. It is likely to lead to slower population growth and reduced GHG emissions in the future. In addition, a better-educated population is likely to be more resilient to the impacts of climate change, independent of population size or growth rate. Investing in female education is therefore another likely candidate for a no-regrets climate policy.

6.3.2 Population policy and the expected costs of climate change: The "greenhouse externality" to childbearing

One way of conceptualizing the role of population-related policies in addressing climate change is through the "greenhouse externality" to childbearing, which we introduced in Chapter 3. Some externalities that arise from childbearing are due to the free-access nature of the global climate resource. Institutional responses that control access are the most direct means of eliminating such externalities. The development of a system of tradable permits allocating rights to GHG emissions is a nascent form of controlling "access" to the global climate. However, population growth can exacerbate free-access problems while institutions are evolving, and Keyfitz (1992) argues that, in a world of imperfect institutions, policies to slow population growth are warranted even if they are not the most direct solution to the problem at hand. In addition, even assuming that the world successfully manages the climate resource by stabilizing GHG concentrations, the difficulty of the task

Box 6.3. How large is the greenhouse externality to childbearing?

A number of studies have estimated the magnitude of the greenhouse externality. Birdsall (1992) considered only LDC births and did not discount future abatement costs, assumed to be US$20 per ton of carbon; her results indicate an externality of about US$1,200 per birth. Cline (1992b) performed a back-of-the-envelope calculation based on a homogeneous, single-region world assuming (based on a cost–benefit model) a 4 gigaton carbon (GtC) global cap on annual carbon emissions. Assuming abatement costs of US$175 per ton carbon, and using discount rates ranging between 2% and 7%, he arrives at an externality of US$700 to US$3,500 per birth averted. Wexler (1996c) improves on the estimates of Cline and Birdsall by accounting for the emissions of the future descendants of a birth, which Birdsall leaves out and Cline underestimates. He also considers emissions of methane and nitrous oxide as well as carbon dioxide; incorporates the age structure of emissions; and examines births in MDCs and LDCs separately. He concludes that the externality ranges between US$1,000 and US$13,000 per birth in LDCs, and between US$3,000 and US$20,000 per birth in MDCs depending mostly on the discount rate employed. Nordhaus and Boyer (1998) provide a detailed theoretical basis for the externality and estimates its magnitude at US$400 to US$3,000 per birth depending on the model used and the climate policy regime in place. Finally, O'Neill and Wexler (1998) estimate the externality assuming the world stabilizes atmospheric carbon dioxide concentrations as discussed in Chapter 1. They find that the externality to childbearing generally ranges between several hundred and several thousand dollars, depending on a number of factors including the timing of emission reductions, the future path of population growth, and the concentration stabilization level. The discount rate and institutional arrangements such as whether or not emissions trading is employed have particularly strong impacts on the externality.

will be determined in part by population growth. Bongaarts *et al.* (1997) calculate that, assuming stabilization of the atmospheric carbon dioxide concentration at 550 parts per million, the required reduction in 2100 of carbon intensity of economic production relative to a "business as usual" scenario would drop from 63% to 33% if the world were to follow the UN low, rather than medium, population path.

Although estimates cover a wide range (see *Box 6.3*), they suggest that the greenhouse externality to childbearing is significant and negative. Furthermore, Birdsall (1994a) compared the implied costs of reducing emissions by averting births with estimates of costs of averting births through education or family planning. She found the two to be comparable, concluding that policies to encourage low fertility are a cost-effective way of reducing GHG emissions and that "increased spending on population programs is likely to be a part of any optimal carbon reduction strategy."

However, it must be kept in mind that the greenhouse externality to childbearing is only one of an entire family of reproductive externalities, some of which are positive. It is the sum of these that is the policy-relevant figure. For example, Lee and Miller (1991) estimate the nonenvironmental externality to childbearing in the United States to be positive US$55,000, nearly three times annual gross national product (GNP) per capita.[3] Most of this reflects the beneficial impact of an additional birth on intergenerational transfer programs related to pensions and health (see Section 3.4). In LDCs, results are varied, with the estimated net externality being either slightly negative relative to per capita GNP (Bangladesh, Mexico, and Kenya) or enormously negative in cases where newborns dilute public assets in the form of natural resource endowments (for example, petroleum for Saudi Arabia, coal for India, and publicly held rain forest for Brazil). The greenhouse externality (and any other environmental externalities) must take its place alongside these other consequences of childbearing. While the results reviewed above demonstrate that environmental externalities are potentially very large, they also indicate that in some cases they might be entirely offset by other externalities.

The GHG externality, when combined with the indivisibility of the global atmosphere, gives rise to an interesting observation regarding international population assistance. When LDCs take measures to lower fertility, for whatever motive, they create an external benefit for future generations in MDCs in the form of lower GHG emissions.[4] To be globally efficient (and equitable as well) population policy should involve transfers of resources from North to South. This strengthens the case for continuation of international population assistance.

6.3.3   Population policy and the precautionary principle

The precautionary principle has been invoked in part because of the complexity of the climate change problem – that is, uncertainties, nonlinearities, and discontinuities in the climate system's response to GHG emissions. As such complexities are inherently scale related, the precautionary principle clearly argues in favor of slowing population growth as a means of reducing risk.

Population growth can have beneficial impacts in the form of economies of scale, induced technological change, etc. However, some of the more dramatic of these effects are rather speculative, especially at the global level. All such effects, moreover, are long term in nature. The precautionary principle suggests, therefore, that these possible impacts of population growth would be an inappropriate and risky basis for policy. Irreversibility is also relevant in this regard. Slow population growth preserves the option of choosing, at a later time, more rapid growth and, as a result, a more populous world. Rapid population growth in the present closes off the option of choosing a less populous world sometime in the future.

6.3.4 Population policy and equity

Poor regions are likely to suffer the greatest impacts of climate change; in this sense, climate change is expected to worsen global inequities compared with the case of an unperturbed climate. In Chapter 5 we concluded that, on balance, slowing population growth is likely to improve the ability of LDCs to respond to climate change impacts. Therefore, population policies can serve an equity goal. From this point of view, policies that not only lower fertility but also improve human capital are likely to be the most effective.

Climate change and population policies also raise the question of intergenerational equity. If it is unfair to subject future generations to the imperfectly understood risks of climate change, and if rapid population growth multiplies such risks, then both climate and population policies can be justified on intergenerational equity grounds. However, it can be argued that investment in general economic and social development would more effectively decrease the vulnerability of at-risk populations than investment in GHG emission reductions or in policies that encourage lower fertility. Indeed, if one is interested in improving the welfare of the worst-off population, this would suggest targeting those living in poverty today (assuming climate change is not so severe as to make the future poor less well-off than today's poor). Climate change mitigation policies that primarily benefit future generations make less sense from this point of view (Schelling, 1995). Many population policies, however, immediately benefit the worst-off members of the population – specifically, poor women and children – even though their climate impacts may take decades to be felt.

6.4 Conclusion

Many population-related policies – such as voluntary family planning and reproductive health programs, and investments in education and primary health care – improve individual welfare among the least well-off members of the present generation. They also tend to lower fertility and slow population growth, reducing GHG emissions in the long run and improving the resilience of vulnerable populations to climate change impacts. Therefore, they easily qualify as no-regrets policies of the sort identified for priority action by the IPCC. The existence of a climate-related external cost to individuals' fertility decisions lends support to such programs, not

only because they assist couples in having the number of childre
also because they tend to lower desired fertility.

In summary:

- Slowing population growth in either LDCs or MDCs is likely to reduce GHG emissions in the long term and ease the measures necessary to stabilize GHG concentrations. Because of the inertia of population growth, GHG emission reductions can be achieved over the next few decades only by reductions in per capita emissions, whether through reductions in consumption, a shift in consumption patterns, or improvements in technology. However, the long-term GHG emission abatement associated with lower population growth can be achieved only by means of lower fertility in the short term. Estimates of the external costs of GHG emissions associated with population growth suggest that the climate-related returns on population policies can be substantial, although impacts on population aging also need to be taken into account.
- Policies to slow population growth in LDCs are likely to improve the resilience of vulnerable populations to climate-related impacts. Since these populations are disproportionately poor, and since high fertility is concentrated at the bottom of the income distribution, equity-based reasoning strengthens the case for population-related policies. The equity case for climate change mitigation policies is weakened by the fact that the benefits will be enjoyed by many future generations, while the poorest generation is probably the current one. In the case of policies related to population, however, the poorest members of the current generation also reap benefits in terms of improved health, human capital, and empowerment.
- MDCs characterized by sub- or nearly sub-replacement fertility are unlikely to display much enthusiasm for policies that would further slow the growth of their populations. However, not just equity, but global efficiency logic, as well, suggests that population policies in LDCs should receive financial support from MDCs.

These conclusions do not necessarily imply that population policies are the most effective or equitable policies for addressing climate change. Throughout our discussion, we have stressed that there are more direct means of reducing GHG emissions and enhancing the functioning of institutions. However, a portfolio approach suggests that policies related to population should be part of a broad range of policies to mitigate and adapt to climate change, and to global environmental change in general, especially given that many of them are win–win strategies.

In this book we have attempted to fill a gap in the scientific and policy consideration of linkages between population and climate change, but much remains to be done. Concern is growing about the impacts of potential changes in the frequency and severity of extreme climate events. If this possibility continues to be a point of focus, more emphasis will need to be placed on how demographic factors affect the vulnerability of populations. Such research to date has concentrated on rural populations, and too little is known about population and environment in urban settings. The international population research agenda is likely to continue to shift from high fertility to population aging, and many links between climate change and population age structure need to be explored. The treatment of population in climate change assessment models can be greatly improved, as can its treatment in global economic models more generally.

The consequences of climate change and demographic change may be substantial in coming decades. Both researchers and policymakers should take into account the linkages between them.

Notes

[1] The impediment is relevant to optimization models for the evaluation of climate change policy, because these models will face this problem if they are to incorporate population as an endogenous variable (Wexler, 1996b).

[2] This conclusion was based on a decomposition of growth rates of variables in the I=PAT equation. As discussed in Chapter 4, such exercises suffer from a number of shortcomings, and this one is no exception – among other problems, it commits the error of failing to disaggregate data, which biases the result toward overestimation of the role of population growth.

[3] As another benchmark, the average middle-income two-parent family in the United States spends US$133,000 to raise a child from birth to age 17 (USDA, 1993).

[4] Following the same logic, if MDCs were to address problems caused by low fertility by means of pronatalist measures, they would impose an external cost on future generations in LDCs. Future generations in MDCs would bear this cost as well, but at least they would enjoy the benefit of stronger pension and health programs.

Appendix I

Fertility and Mortality Assumptions for IIASA Population Projections

Table A.1. Life expectancy at birth, in years.

Region	1995	2000 L	C	H	2030–2035 L	C	H	2080–2085 L	C	H
Male										
North Africa	62.7	63.0	63.8	64.7	64.6	71.1	77.7	64.6	74.9	85.2
Sub-Saharan Africa	50.6	49.6	51.1	52.6	43.1	54.4	65.6	43.1	58.1	73.1
China and CPA	66.4	66.9	67.2	67.4	70.2	72.0	73.9	70.2	75.8	81.4
Pacific Asia	63.1	63.1	64.1	65.1	63.1	70.6	78.1	63.1	74.4	85.6
Pacific OECD	76.1	76.6	77.1	77.6	79.9	83.6	87.4	79.9	87.4	94.9
Central Asia	65.1	65.6	66.1	66.6	68.9	72.6	76.4	68.9	76.4	83.9
Middle East	65.6	65.9	66.7	67.6	67.5	74.0	80.6	67.5	77.8	88.1
South Asia	59.7	59.7	60.5	61.2	59.7	65.3	71.0	59.7	69.1	78.5
Eastern Europe	67.3	67.8	68.3	68.8	71.1	74.8	78.6	71.1	78.6	86.1
European FSU	61.1	61.1	62.1	63.1	61.1	68.6	76.1	61.1	72.4	83.6
Western Europe	72.1	72.6	73.1	73.6	75.9	79.6	83.4	75.9	83.4	90.9
Latin America	66.3	66.8	67.3	67.8	70.1	73.8	77.6	70.1	77.6	85.1
North America	72.3	72.8	73.3	73.8	76.1	79.8	83.6	76.1	83.6	91.1
Female										
North Africa	65.3	65.6	66.4	67.3	67.2	73.7	80.3	67.2	78.7	90.3
Sub-Saharan Africa	53.9	52.9	54.4	55.9	46.4	57.7	68.9	46.4	62.7	78.9
China and CPA	70.1	70.6	71.1	71.6	73.9	77.6	81.4	73.9	82.6	91.4
Pacific Asia	67.4	67.4	68.4	69.4	67.4	74.9	82.4	67.4	79.9	92.4
Pacific OECD	82.2	82.7	83.2	83.7	86.0	89.7	93.5	86.0	94.7	103.5
Central Asia	72.5	73.0	73.5	74.0	76.3	80.0	83.8	76.3	85.0	93.8
Middle East	68.0	68.3	69.1	70.0	69.9	76.4	83.0	69.9	81.4	93.0
South Asia	59.7	59.7	60.7	61.7	59.7	67.2	74.7	59.7	72.2	84.7
Eastern Europe	75.0	75.5	76.0	76.5	78.8	82.5	86.3	78.8	87.5	96.3
European FSU	72.8	73.3	73.8	74.3	76.6	80.3	84.1	76.6	85.3	94.1
Western Europe	78.6	79.1	79.6	80.1	82.4	86.1	89.9	82.4	91.1	99.9
Latin America	71.5	72.0	72.5	73.0	75.3	79.0	82.8	75.3	84.0	92.8
North America	79.1	79.6	80.1	80.6	82.9	86.6	90.4	82.9	91.6	100.4

Abbreviations: L = low; C = central; H = high; CPA = centrally planned Asia; OECD = Organisation for Economic Co-operation and Development; FSU = former Soviet Union.

Table A.2. Total fertility rate.

Region	1995	2000			2030–2035			2080–2085		
		L	C	H	L	C	H	L	C	H
North Africa	4.35	3.92	4.13	4.35	2.00	3.00	4.00	1.54	2.04	2.54
Sub-Saharan Africa	6.18	5.56	5.87	6.18	2.00	3.00	4.00	1.44	1.94	2.44
China and CPA	2.00	1.60	2.00	2.40	1.50	2.25	3.00	1.37	1.87	2.37
Pacific Asia	2.88	2.30	2.65	3.00	1.70	2.35	3.00	1.29	1.79	2.29
Pacific OECD	1.53	1.22	1.53	1.84	1.30	1.70	2.10	1.24	1.74	2.24
Central Asia	3.35	2.68	3.34	4.00	2.00	3.00	4.00	1.45	1.95	2.45
Middle East	5.47	4.92	5.20	5.47	2.00	3.00	4.00	1.45	1.95	2.45
South Asia	3.77	3.39	3.58	3.77	1.70	2.35	3.00	1.20	1.70	2.20
Eastern Europe	1.66	1.33	1.66	2.00	1.30	1.70	2.10	1.39	1.89	2.39
European FSU	1.50	1.20	1.50	1.80	1.30	1.70	2.10	1.55	2.05	2.55
Western Europe	1.67	1.34	1.67	2.00	1.30	1.70	2.10	1.39	1.89	2.39
Latin America	3.10	2.48	2.79	3.10	1.70	2.35	3.00	1.60	2.10	2.60
North America	1.97	1.58	1.94	2.30	1.40	1.85	2.30	1.59	2.09	2.59

Abbreviations: L = low; C = central; H = high; CPA = centrally planned Asia; OECD = Organisation for Economic Co-operation and Development; FSU = former Soviet Union.

Appendix II

Household-level Economies of Scale in Energy Consumption

The first task in formulating a reasonable partitioned model is to determine how much energy in more developed countries (MDCs) and less developed countries (LDCs) is consumed at the level of the household. A portion of this amount is then attributed to "household overhead," that is, that portion of household energy consumption which is independent of the number of members.

MDCs. In MDCs, the residential/commercial sector, the transportation sector, and the industrial sector each account for about one-third of all final energy demand. Direct energy use in homes (i.e., heating, electric appliances, hot water) accounts for 20% of final energy demand or about one-quarter of primary energy (the greater share of primary energy results from the important role of electricity in the household energy supply). If energy for personal transportation (about 15% of primary energy) and energy expended in residential construction and the production of household goods (together about 10% of primary energy) are also assigned to households, it can be estimated that about one-half of all MDC energy use consists of consumption at the level of the household.[1]

Data on US households indicate the presence of strong economies of scale in the residential and transportation sectors. *Table A.3* gives the average household expenditure for direct residential and household vehicle energy consumption (EIA, 1990). These two sectors comprise 20% and 13% of total US final energy demand, respectively. Household energy use is related to household size using both linear and log-linear (constant elasticity) relationships.[2] Both specifications indicate that household overhead (the y-intercept) is 40–50% of average household vehicle energy use and 60–65% of direct residential energy use. The log-linear model implies elasticities of energy use with respect to household size of 0.25 for household vehicles and 0.17 for direct energy, indicating the presence of significant economies of scale.

These parameter estimates should be interpreted with caution, since they confound the effects of household size with those of household income and age structure. However, studies of Australian and Dutch households have found strong evidence of household economies of scale that are not merely a by-product of income or age-structure effects (Ironmonger *et al.*, 1995; Vringer and Blok, 1995).

LDCs. Final energy demand in LDCs is distinguished by reliance on traditional biomass fuels such as animal wastes, crop residues, and wood, or, more compactly,

209

Table A.3. US household energy expenditures for 1988, in US dollars.

Household size (persons)	% of households	Average annual vehicle fuel expenditures	Average residential energy expenditures
1	24.2	577	774
2	33.7	912	1,065
3	16.7	1,189	1,226
4	14.7	1,245	1,320
> 5	10.5	1,422	1,404
Average expenditure		980	1,095
Household overhead			
Linear estimate		410 (40%)	652 (60%)
Log-linear estimate		497 (50%)	701 (64%)

Source: EIA, 1990.

"biofuels." An IIASA–WEC (1995) study estimates that 40% of the energy supply in LDCs comes from biofuels. Nearly all this fuel is used by poor and rural households for cooking and heating. Including modern energy sources used by urban residences, energy consumed at the level of the household appears to account for over half of direct energy consumption even before including household transportation and energy embodied in residential structures and household goods.

Information on household economies of scale, not surprisingly, is more sparse in LDCs than in MDCs. Complicating matters is the fact that economies of scale for modern and traditional fuels differ substantially, and the nature of the fuel basket is closely correlated with household income and household size, as well as other variables (Yousif, 1995). Intuitively, one would expect that household economies of scale for biofuels are at least as great as those for modern fuels. Much of the inefficiency in time and energy from biomass cooking systems results from the necessity of building and maintaining a fire at an appropriate temperature for cooking.[3] All this energy use represents household overhead that is not incurred by modern stoves that deliver energy in more precise and efficient bursts.

Leach (1988) cites household studies in Bangladesh, India, and Pakistan that show per capita energy consumption falling with rising household size; he finds that failure to account for this relationship heavily biases most household energy studies. Fernandez (1980) estimates log-linear models relating energy consumption to household size and income in Pakistan around 1970. The results differ depending on the fuel, but the estimated elasticities of energy consumption with respect to household size (holding income constant) are in the range of 0.2 to 0.4. One study by Cline-Cole *et al.* (1990) in West Africa found that the relationship between household size and total wood fuel consumption was not only highly nonlinear, but nonmonotonic as well. In Jiangsu Province, China, Wang and Feng (1997) found that per capita use of energy for cooking dropped sharply between a household size of 2 and 3, was constant between 3 and 4, and again dropped sharply between 4 and 5, after which it remained relatively constant.

Summary. From the data given above for the MDCs, a case could be made for assigning anywhere from 25–50% of all energy use to household overhead: roughly half of all energy is consumed at the level of the household and, according to the estimates in *Table A.3*, roughly half of this consists of household overhead. Multiplying the two terms together, we arrive at the lower of the two bounds given. The upper bound results from assuming that household-level economies of scale in energy consumption also apply to consumption of other goods. Given economies of scale in food preparation and consumption, and in consumer durables such as furniture, television sets, etc., it seems reasonable to accept the upper-end estimate of 50%. For modern fuel use in LDCs, the 25–50% range is also reasonable. The sectoral decomposition of modern fuel use (in particular, the share of residential energy) is similar in LDCs and MDCs. For biofuels, 50% should be considered a minimum share to be assigned to household overhead. These fuels are used almost exclusively in the residential sector, and economies of scale are greater than in the case of modern fuels.[4] Again, between economies of scale in food and household equipment and the greater economies of scale in traditional energy consumption, the upper-end estimate of 50% appears more likely.

Notes

[1] Authors' calculations based on OECD (1993). The estimates for energy embodied in residential structures and household goods are based on data for the Netherlands (Vringer and Blok, 1995).

[2] Assume E is total household energy, x is household size, and a is household energy overhead. The linear model is $E = bx + a$, where b is energy use per person. The log-linear model is $E = ax^b$, where b is the elasticity of energy use with respect to household size.

[3] Economies of scale in food are analyzed by Deaton and Paxson (1998).

[4] Within the MDC and LDC regions there exists substantial heterogeneity. Among the Organisation for Economic Co-operation and Development countries, for example, the reported share of the residential sector varies from as low as 12% in New Zealand and Spain to as high as 29% in Denmark. Certainly some of this discrepancy is a result of accounting differences, although exactly how much is not clear. Second, patterns of energy use change over time. The IIASA–WEC (1995) report projects that in MDCs the residential/commercial sector will increase from one-third of all energy use to over one-half by 2100, mainly due to the increasing importance of household services and leisure activities. In LDCs, in contrast, the total importance of the residential sector will decrease substantially as biofuels are projected to be phased out by the end of the century.

Appendix III

Population in Major Climate Change Assessment Models

III.1 Projection/Scenario Models

The *Edmonds–Reilly–Barnes Model (ERB)* (Edmonds *et al.*, 1995; Edmonds *et al.*, 1986a; Edmonds and Reilly, 1985) projects energy use by supply technology and associated greenhouse gas (GHG) emissions for nine world regions until the year 2095. The main role of population is its effect on regional gross domestic product (GDP), which is determined by projections of labor force multiplied by exogenous productivity assumptions. Labor force is defined as total population lagged by one time period. Energy demand (E) in the non-Organisation for Economic Co-operation and Development (OECD) countries and the industrial sector of the OECD countries is a function of GDP, per capita income, and energy price:

$$E = P^{\beta_1} y^{\beta_2} Y,$$

where P is an index of energy prices, y is an index of per capita gross national product (GNP), and Y is total GNP. OECD residential and commercial energy demand is a direct function of population:

$$E = P^{\beta_3} y^{\beta_4} * \text{population}.$$

ASF-EPA and *ASF-IPCC (Atmospheric Stabilization Framework)*, emission scenarios used by the US Environmental Protection Agency (EPA) and the Intergovernmental Panel on Climate Change (IPCC), are derived from a core model developed by the EPA called the Atmospheric Stabilization Framework (ASF). The ASF is a linked set of nine regional models of agriculture, energy, and land use. The energy model is a version of ERB (see above) modified to interface with more detailed end-use energy models in the short term. Details about ASF and EPA scenarios can be found in Lashof and Tirpak (1990). Details of the IPCC scenarios and modifications to the model can be found in Pepper *et al.* (1992). In most sectors of the ASF, population simply scales economic activity up or down. In some sectors, such as emissions from human wastes and emissions from agriculture, this leads to a direct linear relationship between emissions and population size. In other sectors, such as emissions from energy use and industry, the relation between population and emissions is also affected by nonlinear relationships between the level of economic demand and the GHG intensity of economic activity. Deforestation assumptions are left unspecified in the model framework. The EPA uses exogenous scenarios, associating a rapid deforestation scenario with rapid population growth and a slow deforestation scenario with slower population growth. The IPCC links deforestation linearly to the size of the population, lagged by 25 years.

The EPA constructs two sets of scenarios: the "slowly changing world," with relatively slow economic growth and technological development, and the "rapidly changing world," with more rapid economic growth and technological development. The slowly changing world is associated with 1987 population projections of the US census that yield a global population of 13.5 billion by 2100. The rapidly changing world is associated with the World Bank's 1988 projections of only 10.5 billion people by 2100, resulting from replacement fertility being reached by 2040 worldwide. The IPCC constructs five different scenarios, with widely varying assumptions about economic growth, deforestation rates, and halogen usage. Three population projections are used. IS92a uses the World Bank's 1991 population projections, which yield a 2100 population of 11.3 billion. IS92c uses the UN medium–low projection (6.4 billion by 2100, based on convergence of fertility at below replacement levels), and IS92e uses UN medium–high (17.6 billion by 2100, based on convergence of fertility at above replacement levels).

IMAGE 2.0 (Alcamo, 1994) is a 13-region integrated economic–atmospheric model developed to evaluate the consequences of climate policies. The model is unique in that it directly links regional economic models with finer-scale grid models of the biosphere. Demand in agriculture, energy use, and industry is projected on a per capita basis for each region and simply scaled up by the assumed population of the region. Aside from this linear effect on the scale of economic activity, population growth assumptions have no effect on the model. The only demographic subtlety in the model is that methane emissions from landfills and carbon emissions from biomass energy are scaled to urban rather than total populations.

III.2 Optimization/Endogenous-Policy Models

Cost–benefit models (Cline and Fankhauser). In 1992, Cline published a study of the economics of global warming (Cline 1992b) centered on a simple global-level cost–benefit model. Population growth, per capita economic growth, and baseline emissions growth are exogenous; the optimal path of investment in GHG abatement is endogenous. Fankhauser (1994) elaborated a similar model in which the exogenous assumptions are allowed to vary stochastically. In these and all optimization models that involve discounting, the main effect of more rapid population growth is to raise the value of the future net benefits which accrue to investment undertaken today, thus encouraging more aggressive GHG emission abatement policies. The authors do not, however, calculate the sensitivity of model results to population assumptions.

DICE (Dynamic Integrated model of Climate and the Economy; Nordhaus, 1994) is a one-region, 12-equation model that models costs and benefits of carbon dioxide (CO_2) emissions from energy and chlorofluorocarbon (CFC) emissions for the next 400 years. A more recent multiregion version (Nordhaus, 1996) is based on the same structure. Population and income are related by the following modified Cobb–Douglas equation, which lies at the heart of the model:

$$Y = AK^{0.25}P^{0.75}\Omega(\mu, E),$$

where Y is GNP, A is a scaling factor representing the state of technology, K is capital, and P is population; Ω represents the impact of global warming abatement costs and damages on output, and is in turn a function of μ, the level of policy-induced emissions controls, and E total emissions. Constrained emissions are determined by the accounting identity

$$E = Y\sigma(1 - \mu),$$

where σ is the GHG emissions intensity of GDP under the zero-abatement case. The results of population sensitivity analysis with DICE are discussed in Chapter 4.

GREEN (General Equilibrium Environmental Model) is a 12-region model developed by the OECD for analyzing the economic costs of policies to control CO_2 emissions from energy. Studies using the model have focused on taxation, trade, and international carbon agreements. A description of the model, along with various results, can be found in OECD (1992). At each time step, equilibrium prices and consumption levels of goods (including energy) are solved by equating demand with production. Energy fuels are included in the balance, resulting in estimates of CO_2 emissions. Population enters into the model via the production function, which exhibits decreasing returns to scale, and household demand. The authors do not, however, present sensitivity analyses involving population.

Global 2100 (Manne and Richels, 1992; Manne *et al.*, 1995) is a five-region macroeconomic model of energy use that computes carbon emissions from energy. Energy demand and supply are divided between various electric and nonelectric technologies and are optimized through time to maximize intertemporal consumption. By constraining carbon emissions in the optimization framework, the model computes the cost of GHG emission policies. Population enters into the Global 2100 model through the exogenous input of labor force into the regional production functions. The production functions are nested CES (constant elasticity of substitution) functions of the following form:

$$Y = (aK^{\alpha\rho}L^{(1-\alpha)\rho} + bE^{\rho})^{1/\rho},$$

where Y is production, K is capital, L is the labor force in efficiency units, E is energy, and α and ρ are elasticities of substitution between the factors. In Global 2100, rising energy prices that limit capital investment or CO_2 limitations that limit the growth of energy inputs or capital inputs have the effect of reducing actual per capita income growth from potential per capita income growth. The effect of a change of population growth in this model is equivalent to the effect of a change in potential per capita productivity. The authors do not, however, present population sensitivity analyses.

References

Abler, D.G., Rodríguez, A.G., and Shortle, J.S., 1998, Labor force growth and the environment in Costa Rica, *Economic Modelling*, **15**:477–499.

Adams, R.M., 1990, *The Wellborn Science: Eugenics in Germany, France, Brazil, and Russia*, Oxford University Press, New York, NY, USA.

Adams, R.M., Rosenzweig, C., Peart, R.M., Ritchie, J.T., McCarl, B.A., Glyer, J.D., Curry, R.B., Jones, J.W., Boote, K.J., and Allen, L.H., Jr., 1990, Global climate change and U.S. agriculture, *Nature*, **345**:219–224.

Adams, R.M., Fleming, R.A., Chang, C-C, McCarl, B.A., and Rosenzweig, C., 1995, A reassessment of the economic effects of global climate change on U.S. agriculture, *Climatic Change*, **30**:147–167.

Agarwal, B., 1994, The gender and environment debate: Lessons from India, in L. Arizpe, M.P. Stone, and D.C. Major, eds, *Population and the Environment: Rethinking the Debate*, Westview Press, Boulder, CO, USA, pp. 97–124.

Ahlburg, D., 1987, The impact of population growth on economic growth in developing nations: The evidence from macroeconomic-demographic models, in D.G. Johnson and R. Lee, eds, *Population Growth and Economic Development: Issues and Evidence*, University of Wisconsin Press, Madison, WI, USA.

Ahlburg, D., 1996, Population growth and poverty, in D. Ahlburg, A. Kelley, and K.O. Mason, eds, *The Impact of Population Growth on Well-Being in Developing Countries*, Springer-Verlag, Berlin, Germany, pp. 219–258.

Ahn, T.S., Knodel, L., Lam, D., and Friedman, J., 1998, Family size and children's education in Vietnam, *Demography*, **35**(1):57–70.

Ahuja, V., 1998, Land degradation, agricultural productivity and common property: Evidence from Côte d'Ivoire, *Environment and Development Economics*, **3**:7–34.

Ainsworth, M., and Over, M., 1994, AIDS and African development, *The World Bank Research Observer*, **9**(2):203–240.

Alan Guttmacher Institute, 1999, *Sharing Responsibility: Women, Society, and Abortion Worldwide*, Alan Guttmacher Institute, New York, NY, USA.

Alberini, A., Cropper, M., Fu, T.-T., Krupnik, A., Liu, J.-T., Shaw, D., and Harrington, W., 1997, Valuing health effects of air pollution in developing countries: The case of Taiwan, *Journal of Environmental Economics and Management*, **34**:107–126.

Alcamo, J., ed., 1994, Image 2.0: *Integrated Modeling of Global Climate Change*, Kluwer Academic Press, Dordrecht, Netherlands.

215

Alcamo, J., Bouman, A., Edmonds, J., and Grübler, A., 1994, An evaluation of the IPCC IS92 emissions scenarios, in J.T. Houghton, L.G. Meira Filho, J. Bruce, H. Lee, B.A. Callander, E. Haites, N. Harris, and K. Maskell, eds, *Climate Change 1994*, Intergovernmental Panel on Climate Change, Cambridge University Press, Cambridge, UK, pp. 250–304.

Alfsen, K.H., Bye, T., Glomsrød, S., and Wiig, H., 1997, Soil degradation and economic development in Ghana, *Environment and Development Economics*, **2**:119–143.

Ali, M., 1995, Institutional and socioeconomic constraints on the second-generation Green Revolution: A case study of Basmati rice production in Pakistan's Punjab, *Economic Development and Cultural Change*, **43**(4):835–861.

Allan, R.J., and D'Arrigo, R.D., 1999, 'Persistent' ENSO sequences: How unusual was the 1990–1995 El Niño?, *The Holocene*, **9**:101–118.

Allen, J., and Barnes, D., 1985, The causes of deforestation in developing countries, *Annals of the Association of American Geographers*, **75**(2):163–184.

Alley, R.B., Meese, D.A., Shuman, C.A., Gow, A.J., Taylor, K.C., Grootes, P.M., White, J.W.C., Ram, M., Waddington, E.D., Mayewski, P.A., and Zielinski, G.A., 1993, Abrupt increase in Greenland snow accumulation at the end of the Younger Dryas event, *Nature*, **362**:527–529.

Alston, J.M., Edwards, G.W., and Freebairn, J.W., 1988, Market distortions and benefits from research, *American Journal of Agricultural Economics*, **70**:281–288.

Amalric, F., 1995, Population growth and the environmental crisis: Beyond the "obvious," in V. Bhaskar and A. Glyn, eds, *The North, the South, and the Environment: Ecological Constraints and the Global Economy*, St. Martin's Press, New York, NY, USA, pp. 85–101.

Anania, G., and McCalla, A.F., 1995, Assessing the impact of agricultural technology improvements in developing countries in the presence of policy distortions, *European Review of Agricultural Economics*, **22**:5–24.

Andersen, L.E., 1996, The causes of deforestation in the Brazilian Amazon, *Journal of Environment and Development*, **5**(3):309–328.

Andersen, O., 1991, Occupational impacts on mortality declines in the Nordic countries, in W. Lutz, ed., *Future Demographic Trends in Europe and North America: What Can We Assume Today?*, Academic Press, London, UK, pp. 1–54.

Anderson, E.W., 1992, The political and strategic significance of water, *Outlook on Agriculture*, **21**(4):247–253.

Ang, B.W., 1993, Sector disaggregation, structural effect and industrial energy use: An approach to analyze the interrelationships, *Energy*, **18**(10):1033–1044.

Ang, B.W., 1995, Decomposition methodology in industrial energy demand analysis, *Energy*, **20**(11):1081–1095.

Antle, J.M., and Heidebrink, G., 1995, Environment and development: Theory and international evidence, *Economic Development and Cultural Change*, **43**(3):603–623.

Appendini, K., and Liverman, D., 1994, Agricultural policy and climate change in Mexico, *Food Policy*, **19**(2):149–164.

Arrhenius, S., 1896, On the influence of carbonic acid in the air upon the temperature of the ground, *The London, Edinburgh, and Dublin Philosophical Magazine and Journal of Science*, **41**:237–276.

Arrow, K.A., Bolin, B., Costanza, R., Dasgupta, P., Folke, C., Holling, C.S., Jansson, B.-O., Levin, S., Mäler, K.-G., Perrings, C., and Pimentel, D., 1995, Economic growth, carrying capacity, and the environment, *Science*, **268**:520–521.

Arrow, K.J., Parikh, J., Pillet, G., Grubb, M., Haites, E., Hourcade, J.C., Parikh, K., and Yamin, F., 1996, Decision-making frameworks for addressing climate change, in J.P. Bruce, H. Lee, and E.F Haites, eds, *Climate Change 1995: Economic and Social Dimensions of Climate Change*, Cambridge University Press, Cambridge, UK, pp. 53–77.

Asante-Duah, D.K., and Sam, P.A., 1995, Assessment of waste management practices in sub-Saharan Africa, *International Journal of Environment and Pollution*, **5**(2/3):224–242.

Ashford, L.S., and Noble, J.A., 1996, Population policy: Consensus and challenges, *Consequences*, **2**(2):25–35.

Asian Development Bank, 1994, *Climate Change in Asia: Indonesia Country Report*, Asian Development Bank, Manila, Philippines.

Attfield, R., 1998, Existence value and use value, *Ecological Economics*, **24**:163–168.

Auerbach, A.J., and Kotlikoff, L.J., 1987, *Dynamic Fiscal Policy*, Cambridge University Press, Cambridge, UK.

Auerbach, A.J., Kotlikoff, L.J., Hagemann, R.P., and Nicoletti, G., 1989, The economic dynamics of an ageing population: The case of four OECD countries, *OECD Economic Studies*, **12**:97–130.

Azhar, R.A., 1993, Commons, regulation and rent-seeking behavior: The dilemma of Pakistan's Guzara forests, *Economic Development and Cultural Change*, **42**(1): 115–229.

Bade, K.J., 1993, Immigration and integration, *European Review*, **1**(1):75–79.

Balk, D., 1994, Individual and community aspects of women's status and fertility in rural Bangladesh, *Population Studies*, **48**(1):21–45.

Banuri, T., Goran-Maler, K., Grubb, M., Jacobson, H.K., and Yamin, F., 1996, Equity and social considerations, in J.P Bruce, H. Lee, and E. Haites, eds, *Climate Change 1995: Economic and Social Dimensions of Climate Change*, Cambridge University Press, Cambridge, UK, pp. 80–124.

Barbier, E.B., 1997, Introduction to the environmental Kuznets curve special interest, *Environment and Development Economics*, **2**:369–381.

Barro, R., 1974, Are government bonds net wealth?, *Journal of Political Economy*, **82**(6):1095–1117.

Bartiaux, F., and van Ypersele, J.-P., 1993, The role of population growth in global warming, in *Proceedings of the International Population Conference Montreal 1994*, Vol. 4, International Union for the Scientific Study of Population, Liège, Belgium.

Beaumont, P., 1994, The myth of Middle Eastern water wars, *Water Resources Development*, **10**(1):9–21.

Becker, G.S., 1981, *A Treatise on the Family*, Harvard University Press, Cambridge, MA, USA.

Becker, G.S., and Barro, R.J., 1988, Reformulating the economic theory of fertility, *Quarterly Journal of Economics*, **103**:1–25.

Becker, G.S., Murphy, K., and Tamura, R., 1991, Human capital, fertility, and economic growth, *Journal of Political Economy*, **98**(5):512–537.

Behrman, J., and Birdsall, N., 1988, The reward for good timing: Cohort effects and earnings functions for Brazilian males, *Review of Economics and Statistics*, **70**(1): 129–135.

Bekki, S., and Law, K.S., 1997, Sensitivity of the atmospheric CH_4 growth rate to global temperature changes observed from 1980 to 1992, *Tellus*, **B49**:409–416.

Bie, S.W., 1990, Dryland degradation measurement techniques, World Bank Environment Department Working Paper 26, The World Bank, Washington, DC, USA.

Bijlsma, L., 1996, Coastal zones and small islands, in R.T. Watson, M.C. Zinyowera, and R.H. Moss, eds, *Climate Change 1995: Impacts, Adaptation, and Mitigation of Climate Change: Scientific-Technical Analyses*, Cambridge University Press, Cambridge, UK.

Bilsborrow, R., 1987, Population pressures and agricultural development in developing countries: A conceptual framework and recent evidence, *World Development*, **15**(2):182–203.

Bilsborrow, R., 1992, Population, development and deforestation: Some recent evidence, *Carolina Population Center Papers*, No. 92–94, University of North Carolina, Chapel Hill, NC, USA.

Bilsborrow, R., 1994, Population, development and deforestation: Some recent evidence, in *Proceedings of the United Nations Expert Group Meeting on Population, Environment, and Development, 1992*, United Nations, New York, NY, USA, pp. 117–134.

Bilsborrow, R., and DeLargy, P., 1991, Land use, migration and natural resource deterioration: The experience of Guatemala and the Sudan, in K. Davis and M. Bernstam, eds, *Resources, Environment and Population*, Oxford University Press, New York, NY, USA, pp. 125–147.

Bilsborrow, R., and Geores, M., 1994, Population, land use and the environment in developing countries: What can we learn from cross-national data?, in D. Pearce and K. Brown, eds, *The Causes of Tropical Deforestation*, University College, London, UK.

Bilsborrow, R., and Okoth-Ogendo, H.W.O., 1992, Population-driven changes in land-use in developing countries, *Ambio*, **21**(1):37–45.

Binswanger, H., 1989, The policy response of agriculture, in *Proceedings of the World Bank Annual Conference on Development Economics*, The World Bank, Washington, DC, USA, pp. 231–258.

Birdsall, N., 1992, Another look at population and global warming, *Population, Health, and Nutrition Policy Research Working Paper WPS 1020*, The World Bank, Washington, DC, USA.

Birdsall, N., 1994a, Another look at population and global warming, in *Population, Environment, and Development*, United Nations, New York, NY, USA, pp. 39–54.

Birdsall, N., 1994b, Government, population, and poverty: A "win–win" tale, in R. Cassen, ed., *Population and Development: Old Debates, New Conclusions*, Transaction Publishers, New Brunswick, NJ, USA, pp. 253–274.

Birdsall, N., and Griffin, C., 1988, Fertility and poverty, *Journal of Policy Modeling*, **10**(1):29–55.

Birdsall, N., and Griffin, C., 1993, Population growth, externalities, and poverty, *Policy Research Working Paper: Population, Health, and Nutrition*, No. WPS 1158, The World Bank, Washington, DC, USA.

Biswas, A.K., 1992, Water for Third World development, *International Journal of Water Resources Development*, **8**(1):3–9.

Biswas, A.K., 1994a, Sustainable water resources development, *International Journal of Water Resources Development*, **10**(2):107–117.

Biswas, A.K., 1994b, Considerations for sustainable irrigation development in Asia, *International Journal of Water Resources Development*, **10**(4):445–455.

Biswas, A.K., 1994c, Management of international water resources: Some recent developments, in A.K. Biswas, ed., *International Waters of the Middle East from Euphrates–Tigris to Nile*, Oxford University Press, Bombay, India.

Biswas, A.K., 1998, Deafness to global water crisis: Causes and risks, *Ambio*, **27**(6): 492–493.

Black, R., 1994, Forced migration and environmental change: The impact of refugees on host environments, *Journal of Environmental Management*, **42**:261–277.

Black, R., and Sessay, M., 1998, Forced migration, natural resource use and environmental change: The case of the Senegal River valley, *International Journal of Population Geography*, **4**:31–47.

Blaikie, P., 1992, Population change and environmental management: Coping and adaptation at the domestic level, in B. Zaba and J. Clarke, eds, *Environment and Population Change*, Ordina Editions, Liège, Belgium, pp. 63–86.

Blanchet, D., 1988, A stochastic version of the Malthusian trap model: Consequences for the empirical relationship between economic growth and population growth in LDCs, *Mathematical Population Studies*, **1**(1):79–99.

Blanchet, D., 1991, Estimating the relationship between population growth and aggregate economic growth in developing countries: Methodological problems, in *Consequences of Rapid Population Growth and Aggregate Economic Growth in Developing Countries, Proceedings of the United Nations/Institut National d'Etudes Démographiques Expert Group Meeting*, New York, 23–26 August 1988, Taylor and Francis, New York, NY, USA.

Bloom, D., and Freeman, R., 1987, Population growth, labor supply, and employment in developing countries, in D.G. Johnson and R.D. Lee, eds, *Population Growth and Economic Development: Issues and Evidence*, University of Wisconsin Press, Madison, WI, USA.

Bloom, D., and Freeman, R., 1988, Economic development and the timing and components of population growth, *Journal of Policy Modeling*, **10**(1):57–82.

Bluffstone, R.A., 1995, The effect of labor market performance on deforestation in developing countries under open access: An example from Nepal, *Journal of Environmental Economics and Management*, **29**:42–63.

Bluffstone, R.A., 1998, Reducing degradation of forests in poor countries when permanent solutions elude us: What instruments do we really have?, *Environment and Development Economics*, **3**:295–317.

Blunier, T., Chappellaz, J., Schwander, J., Dallenbach, A., Stauffer, B., Stocker, T.F., Raynaud, D., Jouzel, J., Clausen, H.B., Hammer, C.U., and Johnsen, S.J., 1998, Asynchrony of Antarctic and Greenland climate change during the last glacial period, *Nature*, **394**:739–743.

Bodansky, D., 1991, Scientific uncertainty and the precautionary principle, *Environment*, **33**(7):4–5, 43–44.

Bodansky, D., 1993, The United Nations Framework Convention on Climate Change: A commentary, *Yale Journal of International Law*, **18**:451–558.

Bok, S., 1994, Population and ethics: Expanding the moral space, in G. Sen, A. Germain, and L.C. Chen, eds, *Population Policies Reconsidered: Health, Empowerment, and Rights*, Harvard University Press, Boston, MA, USA, pp. 15–26.

Bongaarts, J., 1990, The measurement of wanted fertility, *Population and Development Review*, **16**:487.

Bongaarts, J., 1992, Population growth and global warming, *Population and Development Review*, **18**(2):299–319.

Bongaarts, J., 1993, Population growth and the food supply: Conflicting perspectives, *Population Council Working Paper* No. 53, Population Council, New York, NY, USA.

Bongaarts, J., 1994a, The impact of population policies: Comment, *Population and Development Review*, **20**(3): 616–620.

Bongaarts, J., 1994b, Population policy options in the developing world, *Science*, **263**: 771–776.

Bongaarts, J., 1996, Global trends in AIDS mortality, in W. Lutz, ed., *The Future Population of the World: What Can We Assume Today?*, Earthscan, London, UK, pp. 170–195.

Bongaarts, J., and Potter, R., 1983, *Fertility, Biology, and Behavior*, Academic Press, New York, NY, USA.

Bongaarts, J., Mauldin, W.P., and Phillips, J.E., 1990, The demographic impact of family planning programs, *Studies in Family Planning*, **21**(6):299–310.

Bongaarts, J., O'Neill, B.C., and Gaffin, S.R., 1997, Climate change policy: Population left out in the cold, *Environment*, **39**(9):40–41.

Bonneaux, L., 1994, Rwanda: A case of demographic entrapment, *The Lancet*, **344**(December):1689–1690.

Börsch-Supan, A., 1991, The impact of population ageing on savings, investment and growth in the OECD area, in *Future Global Capital Shortages: Real Threat or Pure Fiction?*, OECD, Paris, France.

Boserup, E., 1965, *The Conditions of Agricultural Growth*, Aldine, Chicago, IL, USA.

Boserup, E., 1981, *Population and Technological Change: A Study of Long-Term Trends,* University of Chicago Press, Chicago, IL, USA.

Bouma, M.E., Sondorp, H.E., and van der Kaag, H.J., 1994, Climate change and periodic epidemic malaria, *The Lancet,* **343**:1440.

Bouwman, A.F., 1998, Nitrogen oxides and tropical agriculture, *Nature,* **392**:866–867.

Boyce, J., 1989, Population growth and real wages of agricultural labourers in Bangladesh, *Journal of Development Studies,* **25**(4):467–485.

Bradley, D.J., 1993, Human tropical diseases in a changing environment, *Ciba Foundation Symposium,* **175**:146–162.

Bradley, P.N., and Campbell, B.M., 1998, Who plugged the gap? Re-examining the wood-fuel crisis in Zimbabwe, *Energy and Environment,* **9**(3):235–255.

Brander, J.A., and Taylor, M.S., 1998, The simple economies of Easter Island: A Ricardo-Malthus model of renewable resource use, *American Economic Review,* **88**(1): 119–138.

Broecker, W.S., 1987, Unpleasant surprises in the greenhouse?, *Nature,* **328**:123–126.

Broecker, W.S., 1997, Thermohaline circulation, the Achilles' heel of our climate system: Will man-made CO_2 upset the current balance?, *Science,* **278**:1582–1588.

Brown, H.S., and Eisenberg, H.L., eds, 1995, *The Best Intentions: Unintended Pregnancy and the Wellbeing of Children and Families,* National Academy Press, Washington, DC, USA.

Brown, L., 1985, A false sense of security, in L. Brown, ed., *State of the World 1985,* W.W. Norton, New York, NY, USA, pp. 3–22.

Bruce, J.P, Lee, H., and Haites, E.F., eds, 1996, *Climate Change 1995: Economic and Social Dimensions of Climate Change,* Cambridge University Press, Cambridge, UK.

Bryan, F.O., 1998, Climate drift in a multicentury integration of the NCAR Climate System Model, *Journal of Climate,* **11**:1455–1471.

Bucht, B., 1996, Mortality trends in developing countries: A survey, in W. Lutz, ed., *The Future Population of the World: What Can We Assume Today?,* Earthscan, London, UK, pp. 133–148.

Bufalini, J., Finkelstein, P., and Durman, E., 1989, Air duality, in J.B. Smith and D. Tirpak, eds, *The Potential Effects of Global Climate Change on the United States,* Report EPA-230-05-89-050, US Environmental Protection Agency, Washington, DC, USA.

Buiter, W.H., 1988, Death, birth, productivity growth, and debt neutrality, *Economic Journal,* **98**(June):279–293.

Bumpass, L., 1990, What's happening to the family? Interactions between demographic and institutional change, *Demography,* **27**(4):483–498.

Burgos, J., Curtos de Casas, S., Carcavalla, R., and Galíndey Girón, I., 1994, Global climate change influence in the distribution of some pathogenic complexes (malaria and Chagas disease) in Argentina, *Entomologia e Vectores,* **1**(2):69–78.

Caldwell, J., 1991, Review of Lele and Stone 1989, *Population Studies,* **45**(1):174–175.

Caldwell, J., Reddy, P.H., and Caldwell, P., 1986, Periodic high risk as a cause of fertility decline in a changing rural environment: Survival strategies in the 1980–83 South Indian drought, *Economic Development and Cultural Change*, **34**(4).

Callendar, G.S., 1938, The artificial production of carbon dioxide and its influence on temperature, *Quarterly Journal of the Royal Meteorological Society*, **64**:223–240.

Cameron, J., 1994, The status of the precautionary principle in international law, in T. O'Riordan and J. Cameron, eds, *Interpreting the Precautionary Principle*, Earthscan, London, UK, pp. 262–298.

Cannon, R.C., 1998, The implications of predicted climate change for insect pests in the UK, with emphasis on non-indigenous species, *Global Change Biology*, **4**:785–796.

Cao, H.-X., Mitchell, J.F.B., and Lavery, J.R., 1992, Simulated diurnal range and variability of surface temperature in a global climate model for present and doubled CO_2 climates, *Journal of Climate*, **5**:920–943.

Carson, D.J., 1999, Climate modelling: Achievements and prospects. *Quarterly Journal of the Royal Meteorological Society*, **125**:1–27.

Carter, T.R., Porter, J.H., and Parry, M.L., 1991, Climatic warming and crop potential in Europe: Prospects and uncertainties, *Global Environmental Change*, September, pp. 291–312.

Centers for Disease Control, 1989, Heat-related deaths – Missouri, 1979–1988, *Morbidity and Mortality Weekly Report*, **38**:437–439.

Chagnon, S.A., 1987, An assessment of climate change, water resources, and policy research, *Water International*, **12**:69–76.

Chaibva, S., 1996, Drought, famine, and environmental degradation in Africa, *Ambio*, **25**(3):212–213.

Chapman, D., and Barker, R., 1991, Environmental protection, resource depletion and the sustainability of developing country agriculture, *Economic Development and Cultural Change*, **39**(4):723–737.

Chaudhury, R.H., 1989, Population pressure and its effects on changes in agrarian structure and productivity in rural Bangladesh, in G. Rodgers, ed., *Population Growth and Poverty in Rural South Asia*, Sage Publications, New Delhi, India.

Chen, L.C., Fitzgerald, W.M., and Bates, L., 1995, Women, politics, and global management, *Environment*, **37**(1):4–9, 31–33.

Chen, R.S., and Kates, R., 1994, World food security: Prospects and trends, *Food Policy*, **19**(2):192–208.

Chen, R.S., Bender, W., Kates, R., Messer, E., and Millman, S.R., 1990, *The Hunger Report: 1990*, Brown University, Providence, RI, USA.

Chenery, H., and Syrquin, M., 1975, *Patterns of Development, 1950–1970*, Oxford University Press, Oxford, UK.

Chitale, M.A., 1995, Institutional characteristics for international cooperation in water resources, *Water Resources Development*, **11**(2):113–123.

Chomitz, K.M., and Birdsall, N., 1991, Incentives for small families: Concepts and issues, in *Proceedings of the World Bank Annual Conference on Development Economics 1990*, The World Bank, Washington, DC, USA, pp. 309–349.

Chopra, K., and Gulati, S.C., 1998, Environmental degradation, property rights and population movements: Hypotheses and evidence from Rajasthan (India), *Environment and Development*, **3**:35–57.

Chourasia, L.D., and Tellam, J.H., 1992, Determination of the effect of surface water irrigation on the groundwater chemistry of a hard rock terrain in central India, *Hydrological Sciences*, **37**(4):313–328.

Christy, J.R., and McNider, R.T., 1994, Satellite greenhouse signal, *Nature*, **367**:325.

Clay, D.C., and Vander Haar, J.E., 1993, Patterns of intergenerational support and childbearing in the Third World, *Population Studies*, **47**(1):67–83.

Clay, D.C., Reardon, T., and Kangasniemi, J., 1998, Sustainable intensification in the highland tropics: Rwandan farmers' investments in land conservation and soil fertility, *Economic Development and Cultural Change*, **46**(2):351–377.

Cleaver, K., and Schreiber, D., 1993, The population, agriculture, and environment nexus in sub-Saharan Africa, The World Bank, Washington, DC, USA.

Cleland, J., 1990, Maternal education and child survival: Further evidence and explanations, in J. Caldwell, ed., *What We Know about Health Transition: The Cultural, Social and Behaviourial Determinants of Health*, The Proceedings of an International Workshop, Canberra, May, 1989, Health Transition Series 2, Australian National University, Canberra, Australia.

Cleland, J., 1993, Equity, security and fertility: A reply to Thomas, *Population Studies*, **47**:344–352.

Cleland, J., 1996, A regional review of fertility trends in developing countries: 1960 to 1995, in W. Lutz, ed., *The Future Population of the World: What Can We Assume Today?*, Earthscan, London, UK, pp. 47–72.

Cleland, J., and Rodriguez, G., 1988, The effects of parental education on marital fertility, *Population Studies*, **42**(3):419–442.

Cleland, J., and van Ginnekan, J., 1988, Maternal education and child survival in developing countries: The search for pathways of influence, *Social Science and Medicine*, **27**(12):1357–1368.

Cleland, J., and Wilson, C., 1987, Demand theories of the fertility transition: An iconoclastic view, *Population Studies*, **41**(1):5–30.

Cleveland, J., 1991, Natural resource scarcity and economic growth revisited: Economic and biophysical perspectives, in R. Costanza, ed., *Ecological Economics: The Science and Management of Sustainability*, Columbia University Press, New York, NY, USA, pp. 289–317.

Cline, W.R., 1992a, Global Warming: The Economic Stakes. Policy Analyses in International Economics 36, Institute for International Economics, Washington, DC, USA.

Cline, W.R., 1992b, *The Economics of Global Warming*, Institute for International Economics, Washington, DC, USA.

Cline-Cole, R., Main, H., and Nichol, J., 1990, On fuelwood consumption, population dynamics and deforestation in Africa, *World Development*, **18**(4):513–527.

Coale, A.J., 1973, The demographic transition, in *Proceedings of the International Population Conference*, Vol. 1, International Union for the Scientific Study of Population, Liège, Belgium.

Coale, A.J., and Hoover, E., 1958, *Population Growth and Economic Development in Low-Income Countries*, Princeton University Press, Princeton, NJ, USA.

Cochrane, S., 1979, *Fertility and Education: What Do We Really Know?*, Johns Hopkins University Press, Baltimore, MD, USA.

Cohen, J. 1995, *How Many People Can the Earth Support?*, W.W. Norton, New York, NY, USA.

Colaccio, D., Osborne, T., and Alt, K., 1989, Economic damages from soil erosion, *Journal of Soil and Water Conservation*, **44**:35–29.

Cole, W., and Fayissa, B., 1991, The urban subsistence labour force: Towards a policy-oriented and empirically accessible taxonomy, *World Development*, **19**(7):779–789.

Commoner, B., 1971, *The Closing Circle*, Alfred A. Knopf, New York, NY, USA.

Commoner, B., 1972, Response, *Bulletin of the Atomic Scientists*, May:17, 42–56.

Commoner, B., 1991, Rapid population growth and environmental stress, in *Consequences of Rapid Population Growth in Developing Countries*, Proceedings of the United Nations/Institute d'études démographiques Expert Group Meeting, New York, 23–26 August 1988, Taylor and Francis, New York, NY, USA, pp. 161–190.

Commoner, B., Corr, M., and Stamler, P.J., 1971, The causes of pollution, *Environment*, **13**(3):2–19.

Concepcion, M., 1996, Population policies and family-planning in Southeast Asia, in W. Lutz, ed., *The Future Population of the World: What Can We Assume Today?*, Earthscan, London, UK, pp. 88–101.

Connor, J.M., 1994, Northern America as a precursor of changes in western European food purchasing patterns, *European Review of Agricultural Economics*, **21**(2):155–173.

Conway, D., Krol, M., Alcamo, J., and Hulme, M., 1996, Future availability of water in Egypt: The interaction of global, regional, and basin scale driving forces in the Nile Basin, *Ambio*, **25**(3):336–342.

Cook, R.C., 1962, How many people have ever lived on earth?, *Population Bulletin*, **18**: 1–19.

Cooke, P.A., 1998, Intrahousehold labor allocation responses to environmental scarcity: A case study from the hills of Nepal, *Economic Development and Cultural Change*, **46**(4):807–830.

Corti, S., Molteni, F., and Palmer, T.N., 1999, Signature of recent climate change in frequencies of natural atmospheric circulation regimes, *Nature*, **398**:799–802.

Cotton, P., 1993, Health threat from mosquitoes rises as flood of century finally recedes, *Journal of the American Medical Association*, **270**:685–686.

Council of Europe, 1997, Recent Demographic Developments in Europe, Council of Europe Publishing, Strasbourg, France.

Cramer, J.C., 1998, Population growth and air quality in California, *Demography*, **35**(1):45–56.

Crook, N., 1996, Population and poverty in classical theory: Testing a structural model for India, *Population Studies*, **50**:173–185.

Cropper, M., and Griffiths, C., 1994, The interaction of population growth and environmental quality, *American Economic Review*, **84**(2):250–254.

Cropper, M., and Oates, W., 1992, Environmental economics: A survey, *Journal of Economic Literature*, **30**(2):675–740.

Crosson, P., 1992, Temperate region soil erosion, in V.W. Ruttan, ed., *Sustainable Agriculture and the Environment: Perspectives on Growth and Constraints*, Westview Press, Boulder, CO, USA, pp. 101–112.

Crosson, P., and Anderson, J.R., 1994, Demand and supply: Trends in global agriculture, *Food Policy*, **19**(2):105–119.

Cruz, M.C., Meyer, C.A., Repetto, R., and Woodward, R., 1992, *Population Growth, Poverty and Environmental Stress: Frontier Migration in the Philippines and Costa Rica*, World Resources Institute, Washington, DC, USA.

Cuddington, J.T., Hancock, J.D., and Rogers, C.A., 1994, A dynamic aggregative model of the AIDS epidemic with possible policy implications, *Journal of Policy Modeling*, **16**(5):473–496.

Cutler, D., Poterba, J., Sheiner, L., and Summers, L., 1990, An aging society: Opportunity or challenge?, *Brookings Papers on Economic Activity 1990*:1, Brookings Institution, Washington, DC, USA.

Dahl, C., 1994, A survey of energy demand elasticities for the developing world, *Journal of Energy and Development*, **18**(1):1–47.

Dai, A., Trenberth, K., and Karl, T.R., 1998, Global variations in droughts and wet spells: 1900–1995, *Geophysical Research Letters*, **25**(17):3367–3370.

Daily, G.C., and Ehrlich, P.R., 1992, Population, sustainability and the earth's carrying capacity, *BioScience*, **42**(10):761–771.

Daily, G.C., Ehrlich, P.R., and Ehrlich, A.H., 1994, Optimum human population size, *Population and Environment*, **15**(6):469–475.

Dalton, M., 1997, The welfare bias from omitting climatic variability in economic studies of global warming, *Journal of Environmental Economics and Management*, **33**: 221–239.

Daly, H.E., 1977, *Steady-State Economics: The Economics of Biophysical Equilibrium and Moral Growth*, W.H. Freeman, San Francisco, CA, USA.

Daly, H.E., 1991, Elements of environmental macroeconomics, in R. Costanza, ed., *Ecological Economics: The Science and Management of Sustainability*, Columbia University Press, New York, NY, USA, pp. 32–46.

Daly, H.E., 1992, Allocation, distribution and scale: Towards an economy that is efficient, just and sustainable, *Ecological Economics*, **6**:185–193.

Dansgaard, W., White, J.W.C., and Johnson, S.I., 1989a, The abrupt termination of the Younger Dryas climate event, *Nature*, **339**:532–534.

Dansgaard, W., Johnsen, S.J., Clausen, H.B., Dahljensen, D., Gundestrup, N.S., Hammer, C.U., Hvidberg, D.S., Steffensen, J.P., Sveinbjornsdottir, A.E., Jouzel, J., and Bond, G., 1989b, Evidence for general instability of past climate from a 250-kyr ice-core record, *Nature*, **364**:218–220.

Dasgupta, P.S., 1986, The ethical foundations of population policy, in D.G. Johnson and R.D. Lee, eds, *Population Growth and Economic Development: Issues and Evidence*, The University of Wisconsin Press, Madison, WI, USA, pp. 631–659.

Dasgupta, P.S., 1993, *An Inquiry into Well-Being and Destitution*, Oxford University Press, Oxford, UK.

Dasgupta, P.S., 1995, The population problem: Theory and evidence, *Journal of Economic Literature*, **33**:1879–1902.

Dasgupta, P.S., 1998, Population, consumption, and resources: Ethical issues, *Ecological Economics*, **24**:139–152.

Davis, K., 1954, The world demographic transition, *Annals of the American Academy of Political and Social Science*, **237**:1–11.

Davis, K., 1963, The theory of change and response in modern demographic history, *Population Index*, **29**(4):345–366.

Davis, K., 1991, Population and resources: Fact and interpretation, in K. Davis and M.S. Bernstam, eds, *Resources, Environment and Population: Present Knowledge*, Oxford University Press, Oxford, UK, pp. 1–21.

Day, L.H., 1991, Upper-age longevity in low-mortality countries: A dissenting view, in W. Lutz, ed., *Future Demographic Trends in Europe and North America: What Can We Assume Today?*, Academic Press, London, UK, pp. 117–128.

Deacon, R.T., 1995, Assessing the relationship between government policy and deforestation, *Journal of Environmental Economics and Management*, **28**:1–18.

Deaton, A., and Paxson, C., 1998, Economies of scale, household size, and the demand for food, *Journal of Political Economy*, **106**(5):897–930.

DeCanio, S.J., 1992, International cooperation to avert global warming: Economic growth, carbon pricing, and energy efficiency, *Journal of Environment and Development*, **1**(1):41–62.

De Graff, D.S., Bilsborrow, R.E., and Guilsky, D.K., 1997, Community-level determinants of contraceptive use in the Philippines: A structural analysis, *Demography*, **34**(3):385–398.

de Haen, H., 1991, Environmental consequences of agricultural growth, in S. Vosti, T. Reardon, and W. von Urff, eds, *Agricultural Sustainability, Growth, and Poverty Alleviation: Issues and Policies*, International Food Policy Research Institute, Washington, DC, USA.

Demeny, P., 1986, Population and the invisible hand, *Demography*, **23**(4):473–487.

Denison, E., 1985, *Trends in American Growth, 1929–82*, The Brookings Institution, Washington, DC, USA.

Desai, S., 1995, When are children from large families disadvantaged? Evidence from cross-national analyses, *Population Studies*, **49**:195–210.

Desai, S., and Alva, S., 1998, Maternal education and child health: Is there a strong causal relationship?, *Demography*, **35**(1):71–81.

Dickson, B., Meincke, J., Vassie, I., Jungclaus, J., and Osterhus, S., 1999, Possible predictability in overflow from the Denmark Strait, *Nature*, **397**:243–246.

Dietz, T., and Rosa, E.A., 1997, Effects of population and affluence on CO_2 emissions, *Proceedings of the National Academy of Sciences USA 94*, pp. 175–179.

Dinar, A., and Wolf, A., 1994, International markets for water and potential for regional cooperation: Economic and political perspectives in the Western Middle East, *Economic Development and Cultural Change*, **43**(1):43–66.

Disney, R., 1996, *Can We Afford to Grow Older?*, MIT Press, Cambridge, MA, USA.

Dlugokencky, E.J., Masarie, K.A., Lang, P.M., and Tans, P.P., 1998, Continuing decline in the growth rate of the atmospheric methane burden, *Nature*, **393**:447–450.

Dlugolecki, A.F., 1996, Financial services, in R.T. Watson, M.C. Zinyowera, and R.H. Moss, eds, *Climate Change 1995: Impacts, Adaptations and Mitigation of Climate Change: Scientific-Technical Analyses*, Cambridge University Press, Cambridge, UK, pp. 539–560.

Dobson, A., and Carper, R., 1992, Global warming and potential changes in host-parasite and disease-vector relationships, in R.L. Peters and T.E. Lovejoy, eds, *Global Warming and Biological Diversity*, Yale University Press, New Haven, CT, USA, pp. 200–216.

Donaldson, P.J., 1990, *Nature Against Us: The United States and the World Population Crisis, 1965–1980*, The University of North Carolina Press, Chapel Hill, NC, USA.

Dove, M.R., 1993, A revisionist view of tropical deforestation and development, *Environmental Conservation*, **20**(1):17–24.

Dovers, S.R., and Handmer, J.W., 1995, Ignorance, the precautionary principle, and sustainability, *Ambio*, **24**(2):92–97.

Downing, T.E., 1991, Vulnerability to hunger in Africa: A climate change perspective, *Global Environmental Change*, December:365–380.

Downing, T.E., ed., 1995, *Climate Change and World Food Security*, Springer-Verlag, Berlin, Germany.

Dregne, H., 1988, Desertification of drylands, in P. Unger, T. Sneed, W. Jordan, and R. Jensen, eds, *Challenges in Dryland Agriculture: A Global Perspective*, Proceedings of the International Conference on Dryland Farming, Texas Agricultural Extension Station, Amarillo/Bushland, TX, USA, pp. 610–612.

Dregne, H., 1990, Erosion and soil productivity in Africa, *Journal of Soil and Water Conservation*, **25**:431–436.

du Bois, F., 1994, Water rights and the limits of environmental law, *Journal of Environmental Law*, **6**(1):73–84.

Duchêne, J., and Wunsch, G., 1991, Population aging and the limits to human life, in W. Lutz, ed., *Future Demographic Trends in Europe and North America: What Can We Assume Today?*, Academic Press, London, UK, pp. 27–40.

Duchin, F., 1994, Structural Economics: Toward a Post-normal Science of Ecological Economics, Paper presented at the Biennial Meeting of the International Society for Ecological Economics, San José, Costa Rica, 24–28 October.

Dudgeon, D., 1995, River regulation in southern China: Ecological implications, conservation and environmental management, *Regulated Rivers: Research and Management*, **11**:35–54.

Dufournaud, C.M., Quinn, J.T., Harrington, J.J., Yu, C.C., Abeygumawardena, P., and Franzosa, R., 1995, A model of sustainable extraction of nontimber forest products in subsistence societies, *Environment and Planning*, **A27**:1667–1676.

Duncan, O.D., 1959, Human ecology and population studies, in P.M. Hauser and O.D. Duncan, eds, *The Study of Population*, University of Chicago Press, Chicago, IL, USA.

Duncan, O.D., 1961, From social system to ecosystem, *Social Inquiry*, **31**:140–149.

Dyson, T., 1991, On the demography of South Asian famines, Part II, *Population Studies*, **45**(2):279–297.

Eagleson, P.S., 1994, The evolution of modern hydrology (from watershed to continent in 30 years), *Advances in Water Research*, **17**:3–18.

Earth Negotiations Bulletin, 1994, A summary of the International Conference on Population and Development (ICPD), **6**(39).

Edmonds, J., and Reilly, J., 1985, *Global Energy: Assessing the Future*, Oxford University Press, New York, NY, USA.

Edmonds, J., Reilly, J., Trabalka, J.R., Reichle, D.E., Rind, D., Lebedeff, S., Palutikof, J.P., Wigley, T.M.L., Lough, J.M., Blasing, T.J., Solomon, A.M., Seidel, S., Keyes, D., and Steinberg, M., 1986a, *Future Atmospheric Carbon Dioxide Scenarios and Limitation Strategies*, Noyes Publications, Park Ridge, NJ, USA.

Edmonds, J., Reilly, J., Gardner, R., and Brenkert, A., 1986b, *Uncertainty in Future Global Energy Use and Fossil Fuel CO_2 Emissions 1975 to 2075* (with appendices), United States Department of Energy, Washington, DC, USA.

Edmonds, J., Wise, M., and Barnes, D., 1995, Carbon coalitions: The cost and effectiveness of energy agreements to alter trajectories of atmospheric carbon dioxide emissions, *Energy Policy*, **23**(4/5):309–335.

Egea, N., 1990, Choice of technique revisited: A critical review of the theoretical underpinnings, *World Development*, **18**(11):1445–1456.

Ehrlich, P.R., and Ehrlich, A., 1990, *The Population Explosion*, Simon and Schuster, New York, NY, USA.

Ehrlich, P.R., and Holdren, J., 1970, The people problem, *Saturday Review*, 4 July, pp. 42–43.

Ehrlich, P.R., and Holdren, J., 1971, Impact of population growth, *Science*, **171**:1212–1217.

Ehrlich, P.R., and Holdren, J., 1972, One-dimensional ecology, *Bulletin of the Atomic Scientists*, May:16, 18–27.

Ehrlich, P.R., Ehrlich, A., and Daily, G., 1993, Food security, population and environment, *Population and Development Review*, **19**(1):1–32.

EIA (Energy Information Administration), 1990, Household Vehicles Energy Consumption, Department of Energy, Washington, DC, USA.

El-Hinnawi, E., 1985, *Environmental Refugees*, United Nations Environment Program, Nairobi, Kenya.

Elliott, L., 1996, Environmental conflict: Reviewing the arguments, *Journal of Environment and Development*, **5**(2):149–167.

Emmer, P.C., 1993, Intercontinental migration as a world historical process, *European Review*, **1**(1):67–74.

Engelman, R., 1998, Profiles in Carbon: An Update on Population, Consumption, and CO_2 Emissions, Population Action International, Washington, DC, USA.

Engelman, R., and LeRoy, P., 1995, *Conserving Land: Population and Sustainable Food Production*, Population Action International, Washington, DC, USA.

Epstein, P.R., 1995, Emerging diseases and ecosystem instability: New threats to public health, *American Journal of Public Health*, **85**:168–172.

Evenson, R., 1988, Population growth, infrastructure and real income in North India, in R. Lee, B. Arthur, A. Kelley, and T.N. Srinivasan, eds, *Population, Food and Rural Development*, Oxford University Press, New York, NY, USA.

Evenson, R., 1993, India: Population pressure, technology, infrastructure, capital formation, and rural incomes, in C.L. Jolly and B.B. Torrey, eds, *Population and Land Use in Developing Countries*, National Academy Press, Washington, DC, USA, pp. 70–98.

Fair, R.C., and Dominguez, K.M., 1991, Effects of the changing US age distribution on macroeconomic equations, *American Economic Review*, **81**:1276–1294.

Falkenmark, M., 1989, Middle Eastern hydropolitics: Water scarcity and conflicts in the Middle East, *Ambio*, **18**(6):350–352.

Falkenmark, M., 1994, Population, environment and development: A water perspective, in *Population, Environment and Development: Proceedings of the United Nations Expert Group Meeting on Population, Environment and Development*, United Nations, New York, NY, USA, pp. 99–116.

Falkenmark, M., and Widstrand, C., 1992, Population and water resources: A critical balance, *Population Bulletin*, **47**(3), Population Reference Bureau, Washington, DC, USA.

Fankhauser, S., 1994, The social costs of greenhouse gas emissions: An expected value approach, *The Energy Journal*, **15**(2):157–184.

FAO (Food and Agriculture Organization), 1994, *Water for Life*, FAO, Rome, Italy.

FAO/UNFPA/IIASA, 1982, Potential Population Supporting Capacities of Lands in the Developing World, Report of Project FPA/INT/513, FAO, Rome, Italy.

Farman, J.C., Gardiner, B.G., and Shanklin, J.D., 1985, Large losses of total ozone in Antarctica reveal seasonal ClO_x/NO_x interaction, *Nature*, **315**:207–210.

Farrow, R.A., 1991, Implications of potential global warming on agricultural pests in Australia, *EPPO Bulletin*, **21**:683–696.

Feacham, R., 1994, Health decline in Eastern Europe, *Nature*, **367**:313–314.

Feldman, A., 1990, Environmental degradation and high fertility in sub-Saharan Africa, Morrison Institute for Population and Resource Studies Working Paper No. 36. Stanford University, Stanford, CA, USA.

Fernandez, J., 1980, *Household Energy in Non-OPEC Developing Countries*, Rand Corporation, Santa Monica, CA, USA.

Fields, G., 1990, Labour markets modelling and the urban informal sector: Theory and evidence, in D. Turnham, B. Salme, and A. Schwarz, eds, *The Informal Sector Revisited*, Organisation for Economic Co-operation and Development, Paris, France.

Findley, S.E., 1994, Does drought increase migration? A study of migration from rural Mali during the 1983–1985 drought, *International Migration Review*, **28**(3):539–553.

Fisher, A.C., and Rubio, S.J., 1997, Adjusting to climate change: Implications of increased variability and asymmetric adjustment costs for investment in water reserves, *Journal of Environmental Economics and Management*, **34**:207–227.

Franco, D., and Munzi, T., 1997, Ageing and fiscal policies in the European Union, *European Economy*, **4**:239–388.

Funtowicz, C.O., and Ravetz, J.R., 1994, Uncertainty, complexity and post-normal science, *Environmental Toxicology and Chemistry*, **13**(12):1–5.

Gaffin, S.R., 1998, World population projections for greenhouse gas emissions scenarios: Mitigation and adaptation strategies for global change, in J. Alcamo and N. Nakićenović, guest editors, *Special Issue on Long-term Greenhouse Emission Scenarios and Their Driving Forces*, **3**(2–4):133–170, Kluwer Academic Publishers, Dordrecht, Netherlands.

Gaffin, S.R., and O'Neill, B.C., 1997, Population and global warming with and without CO_2 targets, *Population and Environment*, **18**(4):389–413.

Galloway, P., 1986, Long-term fluctuations in climate and population in the preindustrial era, *Population and Development Review*, **12**(1):1–24.

Garenne, M., 1996, Mortality in sub-Saharan Africa: Trends and prospects, in W. Lutz, ed., *The Future Population of the World: What Can We Assume Today?*, Earthscan, London, UK, pp. 149–169.

Gates, W.L., Henderson-Sellers, A., Boer, G.J., Folland, C.K., Kitoh, A., McAvaney, B.J., Semazzi, F., Smith, N., Weaver, A.J., and Zeng, Q.C., 1996, Climate models – Evaluation, in J.T. Houghton, L.G. Meira Filho, B.A. Callander, N. Harris, A. Kattenberg, and K. Maskell, eds, *Climate Change 1995: The Science of Climate Change*, Cambridge University Press, Cambridge, UK, pp. 229–284.

Gilland, B., 1986, On resources and economic development, *Population and Development Review*, **12**(2):295–305.

Gilland, B., 1988, Population, economic growth, and energy demand, *Population and Development Review*, **14**(2):233–244.

Glantz, M., 1990, Climate variability, climate change and the development process in sub-Saharan Africa, in H.-J. Karpe, D. Otten, and S.C. Trinidale, eds, *Climate and Development: Climatic Change and Variability and the Resulting Social, Economic and Technological Implications*, Springer-Verlag, Heidelberg, Germany, pp. 173–192.

Gleick, P.H., ed., 1993, *Water in Crisis: A Guide to the World's Fresh Water Resources*, Oxford University Press, New York, NY, USA.

Gleick, P.H., 1994, Water and conflict: Fresh water resources and international security, *International Security*, **18**(1):79–112.

Goeller, H.E., and Weinberg, A.M., 1976, The age of substitutability, *Science*, **191**:683.

Goeller, H.E., and Zucker, A., 1984, Infinite resources: The ultimate strategy, *Science*, **223**:456–462.

Goldsmith, J.R., 1986, Three Los Angeles heat waves, in J.R. Goldsmith, ed., *Environmental Epidemiology: Epidemiological Investigation of Community Environmental Health Problems*, CRC Press, Boca Raton, FL, USA, pp. 73–81.

Golini, A., 1998, How low can fertility be? An empirical exploration, *Population and Development Review*, **24**(1):59–74.

González-Villareal, F., and Garduño, H., 1994, Water resources planning and management in Mexico, *Water Resources Development*, **10**(3):239–255.

Goodland, R., 1992, The case that the world has reached limits: More precisely that current throughput growth in the global economy cannot be sustained, *Population and Environment*, **13**(3):167–182.

Goodland, R., 1997, Environmental sustainability in agriculture: Child matters, *Ecological Economics*, **23**:189–200.

Gore, A., 1992, *Earth in the Balance*, Earthscan, London, UK.

Green, C.H., and Tunstall, S.M., 1991, Is the economic evaluation of environmental resources possible?, *Journal of Environmental Management*, **33**(2):123–141.

Gregg, M.B., 1989, The Public Health Consequences of Disasters, U.S. Centers for Disease Control, Atlanta, GA, USA.

Gregory, J.M., and Mitchell, J.F.B., 1997, The climate response to CO_2 of the Hadley Centre coupled AOGCM with and without flux adjustment, *Geophysical Research Letters*, **24**:1943–1946.

Grossman, G.M., and Krueger, A.B., 1995, Economic growth and the environment, *Quarterly Journal of Economics*, May, pp. 353–577.

Grover, B., and Biswas, A.K., 1993, It's time for a World Water Council, *Water International*, **18**:81–83.

Grover, B., and Howarth, D., 1991, Evolving international collaborative arrangements for water supply and sanitation, *Water International*, **16**:146–152.

Grübler, A., 1994, A Comparison of Global and Regional Emissions Scenarios, WP-94-132, International Institute for Applied Systems Analysis, Laxenburg, Austria.

Grübler, A., and Fujii, Y., 1991, Inter-generational and spatial equity issues of carbon accounts, *Energy*, **16**(11/12):1397–1416.

Grübler, A., and Nakićenović, N., 1991, International burden-sharing in greenhouse gas reduction, *Environmental Policy Division*, The World Bank, Washington, DC, USA.

Habakkuk, H.J., 1962, *American and British Technology in the Nineteenth Century*, Cambridge University Press, Cambridge, UK.

Hahn, R.W., and Stavins, R.N., 1999, What Has Kyoto Wrought? The Real Architecture of International Tradable Permit Markets, Discussion Paper 99-30, Resources for the Future, Washington, DC, USA.

Haille, D.G., 1989, Computer simulation of the effects of changes in weather patterns in vector-borne disease transmission, in J.B. Smith and D. Tirpak, eds, *The Potential Effects of Global Climate Change on the United States*, Report EPA-230-05-89-050, US Environmental Protection Agency, Washington, DC, USA, Appendix G.

Haines, A., Epstein, P.R., and McMichael, A.J., 1993, Global health watch: Monitoring the impacts of environmental change, *Lancet*, **342**:1464–1469.

Hall, C.A.S., Cleveland, C.J., and Kaufman, R., 1986, *Energy and Resource Quality: The Ecology of the Economic Process*, John Wiley, New York, NY, USA.

Hall, S.J., and Matson, P.A., 1999, Nitrogen oxide emissions after nitrogen additions in tropical forests, *Nature*, **400**:152–155.

Hammer, J., 1985, Population Growth and Savings in Developing Countries: A Survey, World Bank Staff Working Paper, No. 687, The World Bank, Washington, DC, USA.

Hammitt, J.K., Lempert, R.J., and Schlesinger, M.E., 1992, A sequential-decision strategy for abating climate change, *Nature*, **357**:315–318.

Hansen, J.E., Wilson, H., Sato, M., Ruedy, R., Shah, K., and Hansen, E., 1995, Satellite and surface temperature data at odds?, *Climatic Change*, **30**:103–117.

Hansen, J.E., Sato, M., Lacis, A., Ruedy, R., Tegen, I., and Matthews, E., 1998a, Climate forcings in the industrial era, *Proceedings of the National Academy of Sciences*, **95**:12753–12758.

Hansen, J.E., Sato, M., Ruedy, R., Lacis, A., and Glascoe, J., 1998b, Global climate data and models: A reconciliation, *Science*, **281**:930–932.

Hardin, G., 1968, The tragedy of the commons, *Science*, **162**:1243–1248.

Harrison, P., 1992, *The Third Revolution: Environment, Population and a Sustainable World*, I.D. Tauris and Company in association with Penguin Books, London, UK.

Harrison, P., 1994, Towards a post-Malthusian human ecology, *Human Ecology Review*, **1**(Summer/Autumn):265–276.

Harvey, P.D., 1996, Let's not get carried away with "reproductive health," *Studies in Family Planning*, **27**(5):283–284.

Hayami, Y., and Ruttan, V., 1971, *Agricultural Development: An International Perspective*, Johns Hopkins University Press, Baltimore, MD, USA.

Hayami, Y., and Ruttan, V., 1987, Rapid population growth and agricultural productivity, in D.G. Johnson and R.D. Lee, eds, *Population Growth and Economic Development: Issues and Evidence*, University of Wisconsin Press, Madison, WI, USA, pp. 57–104.

Hayami, Y., and Ruttan, V., 1991, Rapid population growth and technical and institutional change, in G. Tapinos, D. Blanchet, and D.E. Horlacher, eds, *Consequences of Rapid Population Growth in Developing Countries*, Taylor and Francis, New York, NY, USA, pp. 127–157.

Heath, H., and Binswanger, H., 1996, Natural resource degradation effects of poverty and population growth are largely policy-induced: The case of Colombia, *Environment and Development Economics*, **1**:65–83.

Hederra, R., 1987, Environmental sanitation and water supply during floods in Ecuador (1982–83), *Disasters*, **11**:113–116.

Heilig, G.K., 1994, The greenhouse gas methane (CH_4): Sources and sinks, the impact of population growth, possible interventions, *Population and Environment*, **16**(2): 109–137.

Heller, P.S., and Symansky, S., 1997, Implications for Savings of Aging in the Asian "Tigers," IMF Working Paper WP/97/136, International Monetary Fund, Washington, DC, USA.

Henderson-Sellers, A., Zhang, H., Berz, G., Emanuel, K., Gray, W., Landsea, C., Holland, G., Lighthill, J., Shieh, S-L., Webster, P., and McGuffie, K., 1998, Tropical cyclones and global climate change: A post-IPCC assessment, *Bulletin of the American Meteorological Society*, **79**(1):19–38.

Hennessy, K.J., Gregory, J.M., and Mitchell, J.F.B., 1997, Changes in daily precipitation under enhanced greenhouse conditions, *Climate Dynamics*, **13**:667–680.

Henshaw, S.K., 1998, Unintended pregnancy in the United States, *Family Planning Perspective*, **30**:24–29, 46.

Herrera-Basto, E., Prevots, D.R., Zarate, M.L., Silva, J.L., and Amor, J.S., 1992, First reported outbreak of classical dengue fever at 1,700 meters above sea level in Guerrero State, Mexico, *American Journal of Tropical Medicine*, **46**:449–453.

Higgins, M., and Williamson, J.G., 1997, Age structure dynamics in Asia and dependence on foreign capital, *Population and Development Review*, **23**:261–293.

Hill, A., 1990, Demographic responses to food shortages in the Sahel, in G. McNicoll and M. Cain, eds, *Rural Development and Population: Institutions and Policy*, Oxford University Press, New York, NY, USA [supplement to *Population and Development Review*, **15**].

Hobcroft, J., MacDonald, J., and Rutstein, S., 1985, Demographic determinants of infant and early childhood mortality: A comparative analysis, *Population Studies*, **39**: 363–385.

Hodgson, D., 1991, The ideological origins of the Population Association of America, *Population and Development Review*, **17**:1–34.

Hogendorn, J., 1990, *Economic Development*, Second edition, Harper Collins, New York, NY, USA.

Holden, S.T., Shiferaw, B., and Wik, M., 1998, Poverty, market imperfections and time preferences: Of relevance for environmental policy?, *Environment and Development Economics*, **3**:105-130.

Holdren, J., 1991, Population and the energy problem, *Population and Environment*, **12**(3):231–255.

Holling, C.S., 1994, An ecologist view of the Malthusian conflict, in K. Lindahl-Kiessling and H. Landberg, eds, *Population, Economic Development, and the Environment*, Oxford University Press, New York, NY, USA, pp. 79–103.

Homer-Dixon, T.F., 1991, On the threshold: Environmental changes as causes of acute conflict, *International Security*, **16**:2.

Homer-Dixon, T.F., 1994, Environmental scarcities and violent conflict: Evidence from the cases, *International Security*, **19**(1):5–40.

Homer-Dixon, T.F., and Levy, M., 1995, Correspondence: Environment and security, *International Security*, **30**(30):189–198.

Homer-Dixon, T.F., Boutwell, J.H., and Rathjens, G.W., 1993. Environmental change and violent conflict, *Scientific American*, February, pp. 38–45.

Houghton, D., and Houghton, P., 1997, Explaining child nutrition in Vietnam, *Economic Development and Cultural Change*, **45**(3):541–556.

Houghton, J.T., Jenkins, G., and Ephraums, J., eds, 1990, *Climate Change: The IPCC Scientific Assessment*, Cambridge University Press, Cambridge, UK.

Houghton, J.T., Meira Filho, L.G., Callander, B.A., Harris, N., Kattenberg, A., and Maskell, K., eds, 1996, *Climate Change 1995: The Science of Climate Change*, Cambridge University Press, Cambridge, UK.

Houghton, R.A., 1999, The annual net flux of carbon to the atmosphere from changes in land use 1850–1990, *Tellus*, **51B**:298–313.

Hourcade, J.C., 1996, Estimating the costs of mitigating greenhouse gases, in J.P. Bruce, H. Lee, and E.F. Haites, eds, *Climate Change 1995: Economic and Social Dimensions of Climate Change*, Cambridge University Press, Cambridge, UK, pp. 263–196.

House, W., 1987, Labour market differentiation in a developing economy: An example from urban Juba in Southern Sudan, *World Development*, **15**:877–897.

House, W., 1992, Priorities for urban labour market research in anglophone Africa, *Journal of Developing Areas*, **27**(3):49–68.

Howard, P., and Homer-Dixon, T.F., 1995, Environmental Scarcity and Violent Conflict: The Case of Chiapas, Mexico, American Association for the Advancement of Science, Washington, DC, USA.

Howarth, R.B., Schipper, L., Duerr, P.A., and Strom, S., 1991, Manufacturing energy use in 8 OECD countries: Decomposing the impacts of changes in output, industry structure and energy intensity, *Energy Economics*, **13**:135–142.

Hurrell, J.W., 1995, Decadal trends in the North-Atlantic oscillation: Regional temperatures and precipitation, *Science*, **269**:676–679.

IIASA–WEC (International Institute for Applied Systems Analysis and World Energy Council), 1995, *Global Energy Perspectives to 2050 and Beyond*, International Institute for Applied Systems Analysis, Laxenburg, Austria.

Ironmonger, D.S., Aitken, C.K., Erbas, B., 1995, Economies of scale in energy use in adult-only households, *Energy Economics*, **17**:301–310.

Jackson, C., 1995, Questioning synergism: Win–win with women in population and environment policies?, in B. Zaba and J. Clarke, eds, *Environment and Population Change*, Ordina, Liège, Belgium.

Jackson, W.A., 1998, *The Political Economy of Population Ageing*, Edward Elgar, Cheltenham, UK.

Jiminez, E., 1989, Social sector pricing policy revisited, *Proceedings of the World Bank Annual Conference on Development Economics 1989*, The World Bank, Washington, DC, USA.

Johnson, D.G., 1974, *World Food Problems and Prospects*, University of Chicago, Washington, DC, USA.

Jolly, C.L., 1993, Four theories of population change and the environment, *Population and Environment*, **16**(1):6–90.

Jonas, M., Nilsson, S., Shvidenko, A., Stolbovoi, V., Gluck, M., Obersteiner, M., and Öskog, A., 1999, Full carbon accounting and the Kyoto Protocol: A systems-analytical view, IR-99-025, International Institute for Applied Systems Analysis, Laxenburg, Austria.

Jonas, P.R., Charlson, R.J., and Rodhe, H., 1995, Aerosols, in J.T. Houghton, L.G. Meira Filho, J. Bruce, H. Lee, B.A. Callander, E. Haites, N. Harris, and K. Maskell, eds, *Climate Change 1994: Radiative Forcing of Climate Change and an Evaluation of the IPCC IS92 Emission Scenarios*, Cambridge University Press, Cambridge, UK, pp. 127–162.

Jones, G., 1990, *Population Dynamics and Educational and Health Planning*, International Labour Organization, Geneva, Switzerland.

Jonish, J., 1992, Sustainable Development and Employment: Forestry in Malaysia, World Employment Programme Working Paper WEP-2-22/WP.234, International Labour Organization, Geneva, Switzerland.

Kabubi, J.N., Gellens, D., and Demarée, G.R., 1995, Response of an upland equatorial African basin to a CO_2-induced climate change, *Hydrological Sciences*, **40**(4): 453–470.

Kaczmarek, Z., 1996, Water resources management, in R.T. Watson, M.C. Zinyowera, and R.H. Moss, eds, *Climate Change 1995: Impacts, Adaptations and Mitigation of Climate Change: Scientific-Technical Analyses*, Cambridge University Press, Cambridge, UK, pp. 469–486.

Kaczmarek, Z., Niestepski, M., and Osuch, M., 1995, Climate Change Impact on Water Availability and Use, WP-95-48, International Institute for Applied Systems Analysis, Laxenburg, Austria.

Kaczmarek, Z., Liszewska, M., and Osuch, M., 1997, Water management in South Asia in the 21st century, *Geographica Polonica*, **70**:7–24.

Kalkenstein, L.S., 1989, The impact of CO_2 and trace gas-induced climate change upon human mortality, in J.B. Smith and D. Tirpak, eds, *The Potential Effects of Global Climate Change on the United States*, Report EPA-230-05-89-050, US Environmental Protection Agency, Washington, DC, USA, Appendix G.

Kalkenstein, L.S., 1993a, Global warming and human health, in L. Kalkenstein and J. Smith, eds, *International Implications of Global Warming*, Cambridge University Press, Cambridge, UK.

Kalkenstein, L.S., 1993b, Health and climate change: Direct impacts in cities, *the Lancet*, **342**:1397–1399.

Kalkenstein, L.S., and Smoyer, K.E., 1993, The impact of climate change on human health: Some international implications, *Experientia*, **49**:469–479.

Kane, S., Reilly, J., and Tobey, J., 1992, An empirical study on the economic effects of climate change on world agriculture, *Climatic Change*, **21**:17–35.

Karl, T.R., and Häberli, W., 1998, Climate extremes and natural disasters: Trends and loss reduction prospects, *Proceedings of the Conference on the World Climate Research Programme: Achievements, Benefits, and Challenges*, Geneva, 26–28 August 1997, pp. 166–178, World Meteorological Organization, Geneva, Switzerland.

Karl, T.R., and Knight, R.W., 1998, Secular trends of precipitation amount, frequency, and intensity in the United States, *Bulletin of the American Meteorological Society*, **79**(2):231–241.

Kasting, J.F., Toon, O.B., and Pollack, J.B., 1988, How climate evolved on the terrestrial planets, *Scientific American*, February:46–53.

Kattenberg, A., Georgi, F., Grassl, H., Meeke, G.A., Mitchell, J.F.B., Stouffer, R.J., Tokioka, T., Weaver, A.J., and Wigley, T.M.L., 1996, Climate modes – Projections of future climate, in J.T. Houghton, L.G. Meira Filho, B.A. Callander, N. Harris, A. Kattenberg, and K. Maskell, eds, *Climate Change 1995: The Science of Climate Change*, Cambridge University Press, Cambridge, UK, pp. 285–358.

Katz, R.W., and Brown, B.G., 1992, Extreme events in a changing climate: Variability is more important than averages, *Climate Change*, **21**:289–302.

Katz, R.W., and Brown, B.G., 1994, Sensitivity of extreme events to climate change: The case of autocorrelated time series, *Environmetrics*, **5**:451–462.

Kavalanekar, N.B., Sharma, S.C., and Rushton, K.R., 1992, Over-exploitation of an alluvial aquifer in Gujarat, India, *Hydrological Sciences*, **37**(4):329–346.

Kawashima, H., Bazin, M.J., and Lynch, J.M., 1997, A modelling study of world protein supply and nitrogen fertilizer demand in the 21st century, *Environmental Conservation*, **24**(1):50–56.

Keeling, C.D., 1994, Global historical CO_2 emissions, in *Trends: A Compendium of Data on Global Change*, Carbon Dioxide Information Analysis Center, Oak Ridge National Laboratory, Oak Ridge, TN, USA.

Keeling, C.D., and Whorf, T.P., 1996, Atmospheric CO_2 records from sites in the SIO air sampling network, in *Trends: A Compendium of Data on Global Change*, Carbon Dioxide Information Analysis Center, Oak Ridge National Laboratory, Oak Ridge, TN, USA.

Kelley, A., 1996, The consequences of rapid population growth on human resource development: The case of education, in D. Ahlburg, A. Kelley, and K.O. Mason, eds, *The Impact of Population Growth on Well-Being in Developing Countries*, Springer-Verlag, Berlin, Germany, pp. 67–137.

Kelley, A., and McGreevey, W., 1994, Population and development in historical perspective, in R. Cassen, ed., *Population and Development: Old Debates, New Conclusions*, Overseas Development Council, Washington, DC, USA.

Kelley, A., and Schmidt, R., 1996, Toward a cure for the myopia and tunnel vision of the population debate, in D. Ahlburg, A. Kelley, and K.O. Mason, eds, *The Impact of Population Growth on Well-Being in Developing Countries*, Springer-Verlag, Berlin, Germany, pp. 11–35.

Kellogg, W.W., and Schware, R., 1981, *Climate Change and Society: Consequences of Increasing Atmospheric Carbon Dioxide*, Westview Press, Boulder, CO, USA.

Kelly, D., and Kolstad, C., 1996, Malthus and Climate Change: Betting on a Stable Population, draft manuscript.

Kelly, K., and Homer-Dixon, T.F., 1995, *Environmental Scarcity and Violent Conflict: The Case of Gaza*, American Association for the Advancement of Science, Washington, DC, USA.

Kennedy, V.S., 1990, Anticipated effects of climate changes on estuarine and coastal fisheries, *Fisheries*, **15**(6):16–24.

Kerr, R.A., 1996, A new dawn for sun-climate links?, *Science*, **271**:1360–1361.

Kerr, R.A., 1997, Climate change: Model gets it right – without fudge factors, *Science*, **276**:1041.

Kerr, R.A., 1999, Big El Niños ride the back of slower climate change, *Science*, **283**: 1108–1109.

Keyfitz, N., 1992, Seven ways of making the less developed countries' population problem disappear – In theory, *European Journal of Population*, **8**:149–167.

Khan, A., 1988, Population growth and access to land: An Asian perspective, in R. Lee, B. Arthur, A. Kilbey, G. Rodgers, and T.N. Srinivasan, eds, *Population, Food and Rural Development*, Oxford University Press, New York, NY, USA.

Kinsella, K., and Gist, Y.J., 1995, Older Workers, Retirement, and Pensions: A Comparative International Chartbook, Report No. IPC/95-2RP, Bureau of the Census, Washington, DC, USA.

Kirschbaum, M.U.F., and Fischlin, A., 1996, Climate change impacts on forests, in R.T. Watson, M.C. Zinyowera, and R.H. Moss, eds, *Climate Change 1995: Impacts, Adaptations and Mitigation of Climate Change: Scientific-Technical Analyses*, Cambridge University Press, Cambridge, UK, pp. 95–129.

Kneese, A., 1988, The economics of natural resources, in M.S. Teitelbaum and J.M. Winter, eds, *Population and Resources in Western Intellectual Tradition*, Cambridge University Press, New York, NY, USA, pp. 281–309.

Knowles, J.C., Akin, J.S., and Guilsky, D.K., 1994, The impact of population policies: Comment, *Population and Development Review*, **20**(3):611–615.

Knutson, T.R., Manabe, S., and Gu, D.F., 1997, Simulated ENSO in a global coupled ocean-atmosphere model: Multi-decadal amplitude modulation and CO_2 sensitivity, *Journal of Climate*, **10**:138–161.

Knutson, T.R., Tuleya, R.E., and Kurihara, Y., 1998, Simulated increase of hurricane intensities in a CO_2-warmed climate, *Science*, **279**:1018–1020.

Kohl, R., and O'Brian, P., 1998, The Macroeconomics of Ageing, Pensions and Savings: A Survey, OECD Ageing Working Paper AWP 1.1, OECD, Paris, France.

Kohler, H.-P., 1997, Learning in social networks and contraceptive choice, *Demography*, **34**(3):369–383.

Kolsrud, G., and Torrey, B., 1992, The importance of population growth in future commercial energy consumption, in J. White, ed., *Global Climate Change*, Plenum Press, New York, NY, USA.

Komlos, J., and Artzrouni, M., 1990, Mathematical investigations of the escape from the Malthusian trap, *Mathematical Population Sciences*, **24**(4):269–287.

Koopman, J.S., Prevots, D.R., and Marin, M.A.V., 1991, Determinants and predictors of dengue infection in Mexico, *American Journal of Epidemiology*, **133**:1168–1178.

Kühl, S., 1997, *Die Internationale der Rassisten: Aufstieg und Niedergang der internationalen Bewegung für Eugenik und Rassenhygiene im 20. Jahrhundert*, Campus-Verlag, Frankfurt am Main, Germany.

Kuznets, S., 1966, *Modern Economic Growth*, Yale University Press, New Haven, CT, USA.

Langford, I.H., and Bentham, G., 1993, The Potential Effects of Climate Change on Winter Mortality in the UK, Working Paper GEC 93-25, Centre of Social and Economic Research on the Global Environment (CSERGE), University of East Anglia and University College London.

Lashof, D.A., 1989, The dynamic greenhouse: Feedback processes that may influence future concentration of atmospheric gases and climatic change, *Climatic Change*, **14**(3):213–242.

Lashof, D.A., and Tirpak, D., eds, 1990, *Policy Options for Stabilizing Global Climate*, Hemisphere Publishing Corp., Washington, DC, USA.

Last, J.M., 1993, Global change: Ozone depletion, greenhouse warming and public health, *Annual Review of Public Health*, **14**:115–136.

Law, K.S., and Nisbet, E.G., 1996, Sensitivity of the CH_4 rate to changes in CH_4 emissions from natural gas and coal, *Journal of Geophysical Research*, **101**(D9):14387–14397.

Layard, R., Blanchard, O., Dornbusch, R., and Krugman, P., 1992, *East-West Migration: The Alternatives*, MIT Press, Cambridge, MA, USA.

Leach, G., 1988, Residential energy in the Third World, *Annual Review of Energy*, **V**(13):47–65.

Lee, R., 1980, A historical perspective on economic aspects of the population explosion: The case of pre-industrial England, in R. Easterlin, ed., *Population and Economic Change in Developing Countries*, University of Chicago Press, Chicago, IL, USA, pp. 517–557, 563–566.

Lee, R., 1983, Economic consequences of population size, structure, and growth, *IUSSP Newsletter*, **17**:43–59.

Lee, R., 1987, Population dynamics of humans and other animals, *Demography*, **24**(4):443–465.

Lee, R., 1989, Evaluating externalities to childbearing in developing countries: The case of India, in *Collected Papers from the UN-INED Conference on Population and Development*, New York, NY, USA, August 1988.

Lee, R., 1990, The demographic response to economic crisis in historical and contemporary populations, *Population Bulletin of the United Nations*, **29**:1–15.

Lee, R., 1991, Comment: The second tragedy of the commons, in M. Bernstam and K. Davis, eds, *Resources, Environment, and Population: Present Knowledge, Future Options*, Oxford University Press, New York, NY, USA.

Lee, R., and Miller, T., 1991, Population growth, externalities to childbearing, and fertility policy in the developing world, in *Proceedings of the World Bank Annual Conference on Development Economics, 1990*, The World Bank, Washington, DC, USA, pp. 275–304.

Lee, R., Mason, A., and Miller, T., 1999, Saving, Wealth, and Population, paper presented at Annual Meeting of the Population Association of America, 25–27 March 1999, New York, NY, USA.

Lee, S.-W., 1997, Not a one-time event: Environmental change, ethnic rivalry, and violent conflict in the Third World, *Journal of Environment and Development*, **6**(4): 365–396.

Leggett, J., Pepper, W.J., and Swart, R.J., 1992, Emissions scenarios for the IPCC: An update, in J.T. Houghton, B.A. Callander, and S.K. Varney, eds, *Climate Change 1992: The Supplementary Report to the IPCC Scientific Assessment*, Cambridge University Press, Cambridge, UK, pp. 68–95.

Le Grand, R., and Phillips, L.F., 1996, The effect of fertility reduction on infant and child mortality: Evidence from Matlab in rural Bangladesh, *Population Studies*, **50**:51–68.

Leibenstein, H., 1967, The impact of population growth on "non-economic" determinants of economic growth, in *Proceedings of the World Population Conference, Belgrade, 1965*, United Nations, New York, NY, USA.

Leibenstein, H., 1976, Population growth and savings, in L. Tabah, ed., *Population Growth and Economic Development in the Third World*, Ordina Editions, Dolhain, Belgium.

Lele, U., and Stone, S.W., 1989, Population pressure, the environment and agricultural intensification: Variations on the Boserup hypothesis, *MADIA Discussion Paper 4*, The World Bank, Washington, DC, USA.

Leslie, P.H., 1945, On the use of matrices in certain population mathematics, *Biometrika*, **33**:183–212.

Levins, R., 1995, Preparing for uncertainty, *Ecosystem Health*, **1**:47–57.

Levy, M., 1995a, Time for a third wave of environment and security scholarship?, *Environmental Change and Security Project: Report, Issue 1*, Woodrow Wilson Center, Princeton, NJ, USA, pp. 44–46.

Levy, M., 1995b, Is the environment a national security issue?, *International Security*, **20**(2):35–62.

Liebscher, H.J., 1993, Hydrology for the management of large river basins, *Hydrological Sciences*, **38**(1/2):1–13.

Lindblade, K.A., Carswell, G., and Tumushairwe, J.K., 1998, Mitigating the relationship between population growth and land degradation, *Ambio*, **27**(7):565–571.

Lloyd, C., 1994, Investing in the next generation: The implications of high fertility at the level of the family, in R. Cassen, ed., *Population and Development: Old Debates, New Conclusions*, Overseas Development Council, Washington, DC, USA.

Lockwood, M., 1997, Sons of the soil? Population growth, environmental change and men's reproductive intentions in northern Nigeria, *International Journal of Population Geography*, **3**:305–322.

Loevinsohn, M., 1994, Climate warming and increased malaria incidence in Rwanda, *Lancet*, **343**:714–718.

Lonergan, S., 1998, The role of environmental degradation in population displacement, *Environmental Change and Security Project Report*, Issue 4(Spring):5–15, The Woodrow Wilson Center, Washington, DC, USA.

Lonergan, S., and Kavanaugh, B., 1991, Climate change, water resources, and security in the Middle East, *Global Environmental Change*, September:272–290.

Longstreth, J.D., 1990, Global warming: Clues to potential health effects, *Environmental Carcinogenesis Reviews*, **C8**(1):139–169.

Love, L., and Shewliakova, E., 1998, Potential losses from climate change in the U.S., *Energy and Environment*, **9**(4):347–361.

Lovins, A.B., and Lovins, L.H., 1997, *Climate: Making Sense and Making Money*, Rocky Mountain Institute, Old Snowmass, CO, USA.

Ludwig, H.F., and Islam, A., 1992, Environmental management of industrial wastes in developing countries: Bangladesh case study, *International Journal of Environment and Pollution*, **2**(1/2):76–86.

Lutz, W., 1993, Population and environment – What do we need more urgently: Better data, better models, or better questions?, in B. Zaba and J. Clarke, eds, *Environment and Population Change*, International Union for the Scientific Study of Population, Derouaux Ordina Editions, Liège, Belgium, pp. 47–62.

Lutz, W., ed., 1994, *Population-Development-Environment: Understanding their Interactions in Mauritius*, Springer-Verlag, Berlin, Germany.

Lutz, W., ed., 1996, *The Future Population of the World: What Can We Assume Today?*, Earthscan, London, UK.

Lutz, W., 1997, Determinants and Consequences of the World's Most Rapid Fertility Decline on the Island of Mauritius, Paper presented at the meeting of the Population Association of America, 27–29 March, Washington, DC, USA.

Lutz, W., Sanderson, W., and Scherbov, S., 1996a, Probabilistic population projections based on expert opinion, in W. Lutz, ed., *The Future Population of the World: What Can We Assume Today?*, Earthscan, London, UK, pp. 397–428.

Lutz, W., Folan, W., Gunn, J., and Faust, B., 1996b, Possible Effects of Climate Change on the Classic Maya Collapse, Paper presented at the Population Association of America (PAA) Annual Meeting, 9–11 May 1996, New Orleans, LA, USA.

Lutz, W., Sanderson, W., and Scherbov, S., 1997, Doubling of world population unlikely, *Nature*, **387**:803–805.

MacKellar, L., 1994, Population and development: Assessment in advance of the 1994 World Population Conference, *Development Policy Review*, **12**(June):1–28.

MacKellar, L., and McGreevey, W., 1999, The growth and containment of social security systems, *Development Policy Review*, **17**(1):1–24.

MacKellar, L., and Reisen, H., 1998, A simulation model of global pension investment, *OECD Ageing Working Paper AWP 1.5*, OECD, Paris, France.

MacKellar, L., and Vining, D.R., Jr., 1987, Natural resource scarcity: A global survey, in G. Johnson and R. Lee, eds, *Population Growth and Economic Development: Issues and Evidence*, University of Wisconsin Press, Madison, WI, USA, pp. 259–327.

MacKellar, L., and Vining, D.R., Jr., 1988, Where does the United States stand in the global resource scarcity debate?, *Environment and Planning*, **A20**:1567–1573.

MacKellar, L., and Vining, D.R., Jr., 1989, Measuring natural resource scarcity, *Social Indicators Research*, **21**:517–530.

MacKellar, L., and Vining, D.R., Jr., 1995, Population concentration in less developed countries (LDCs): New evidence, *Papers in Regional Science*, **74**(3):259–293.

MacKellar, L., Lutz, W., Prinz, C., and Goujon, A., 1995, Population, households, and CO_2 emissions, *Population and Development Review*, **21**(4):849–865.

Maddison, J.M., 1989, *The World Economy in the 20th Century*, Organisation for Economic Co-operation and Development, Paris, France.

Magadza, C.H.D., 1994, Climate change: Some likely multiple impacts in southern Africa, *Food Policy*, **19**(2):165–191.

Mageed, Y.A., and White, G.F., 1995, Critical analysis of existing institutional arrangements, *Water Resources Development*, **11**(2):103–111.

Malingreau, J.-P., and Tucker, C.J., 1988, Large-scale deforestation in the southeastern Amazon basin of Brazil, *Ambio*, **17**:49–55.

Malla, P.B., and Gopalakrishan, C., 1995, Conservation effects of irrigation water pricing: A case study from Oahu, Hawaii, *Water Resources Development*, **11**(3):233–242.

Malthus, R., 1798 [1967], *Essay on the Principle of Population*, 7th edition, Dent, London, UK.

Manabe, S., and Stouffer, R.J., 1994, Century-scale effects of increased atmospheric CO_2 on the ocean-atmosphere system, *Nature*, **364**:215–218.

Manabe, S., and Wetherald, R.T., 1967, Thermal equilibrium of the atmosphere with a given distribution of relative humidity, *Journal of Atmospheric Science*, **24**:241–259.

Manabe, S., and Wetherald, R.T., 1975, The effects of doubling the CO_2 concentration on the climate of a general circulation model, *Journal of Atmospheric Science*, **32**:3–15.

Mann, K.H., and Drinkwater, K.F., 1994, Environmental influences on fish and shellfish production in the Northwest Atlantic, *Environmental Review*, **2**:16–32.

Mann, M.E., Bradley, R.S., Hughes, M.K., 1998, Global-scale temperature patterns and climate forcing over the past six centuries, *Nature*, **392**:779–787.

Mann, M.E., Bradley, R.S., Hughes, M.K., 1999, Northern hemisphere temperatures during the past millennium: Inferences, uncertainties, and limitations, *Geophysical Research Letters*, **26**:759–762.

Manne, A., and Richels, R., 1992, *Buying Greenhouse Insurance: The Economic Costs of CO_2 Emissions Limits*, MIT Press, Cambridge, MA, USA.

Manne, A., Richels, R., and Mendelsohn, R., 1995, MERGE: A model for evaluating regional and global effects of GHG reduction policies, *Energy Policy*, **23**(1):17–34.

Manton, K.G., 1991, New biotechnologies and the limits to life expectancy, in W. Lutz, ed., *Future Demographic Trends in Europe and North America: What Can We Assume Today?*, Academic Press, London, UK, pp. 97–115.

Manton, K.G., Stallard, E., and Tolley, H.D., 1991, Limits to human life expectancy, *Population and Development Review*, **17**(4):603–637.

Marden, P.G., and Hodgson, D., 1975, *Population, Environment, and the Quality of Life*, AMS Press, Inc., New York, NY, USA.

Markandya, A., 1998, Poverty, income distribution and policy making, *Environmental and Resource Economics*, **11**(3–4):459–472.

Marland, G., Boden, T.A., Andres, R.J., Brenkert, A.L., and Johnston, C., 1999, Global, regional, and national CO_2 emissions, in *Trends: A Compendium of Data on Global Change*, Carbon Dioxide Information Analysis Center, Oak Ridge National Laboratory, Oak Ridge, TN, USA.

Marquette, C.M., 1995, Population and environment in industrialized regions: Some general policy recommendations, in A. Potrykowska and J.I. Clarke, *Population and Environment in Industrialized Regions*, Polish Academy of Sciences, Warsaw, Poland, pp. 283–298.

Martens, W.J.M., Rotmans, P., and Niessen, L.W., 1994, Climate change and malaria risk: An integrated modelling approach, GLOBO Report Series 3, Report 461502003, Global Dynamics and Sustainable Development Program, National Institute of Public Health and Environmental Protection (RIVM), Bilthoeven, Netherlands.

Martens, W.J.M., Niessen, L., Rotmans, J., Jeten, T.H., and McMichael, A.J., 1995a, Potential impact of global climate change on malaria risk, *Environmental Health Perspectives*, **103**:458–464.

Martens, W.J.M., Retten, T.H., Rotmans, J., and Nielsson, L.W., 1995b, Climate change and vector-borne diseases: A global modelling perspective, *Global Environmental Change*, **5**(3):195–209.

Martin, P.H., and Lefebvre, M.G., 1995, Malaria and climate: Sensitivity of malaria potential transition to climate, *Ambio*, **24**(4):200–207.

Mason, A., 1988, Saving, economic growth and demographic change, *Population and Development Review*, **10**(2):177–240.

Mason, K.O., 1997, Explaining fertility transitions, *Demography*, **34**(4):443–454.

Masson, P., and Tryon, R., 1990, Macroeconomic effects of projected population aging in industrial countries, *IMF Staff Papers*, **37**(3):453–485, International Monetary Fund, Washington, DC, USA.

Mather, A.S., 1989, Global trends in forest resources, in K. Davis and M.S. Bernstam, eds, *Resources, Environment and Population: Present Knowledge*, Oxford University Press, Oxford, UK, pp. 289–304.

Matyasovsky, I., Bogardi, I., Bardossy, A., and Dickstein, L., 1993, Space-time precipitation reflecting climate change, *Hydrological Sciences*, **36**(6):539–558.

Maytín, C.E., Acevedo, M.F., Jaimez, R., Andressen, R., Harwell, M.A., Robock, A., and Azócar, A., 1995, Potential effects of global climatic change on the phenology and yield of maize in Venezuela, *Climate Change*, **29**:189–211.

McCabe, G.J., Jr., and Hay, L.E., 1995, Hydrological effects of hypothetical climate change in the East River basin, Colorado, USA, *Hydrological Sciences*, **40**(3):303–318.

McGreevey, W.P., 1990, Social Security in Latin America, *World Bank Discussion Paper No. 110*, The World Bank, Washington, DC, USA.

McIntosh, C.A., and Finkle, J.L., 1995, The Cairo Conference on Population and Development, *Population and Development Review*, **21**(2):223–260.

McKeown, T.E., 1988, *The Origins of Disease*, Blackwell, Oxford, UK.

McMichael, A.J., 1993, Global environmental change and human population health: A conceptual and scientific challenge for epidemiology, *International Journal of Epidemiology*, **22**:1–8.

McMichael, A.J., 1996, Human population health, in R.T. Watson, M.C. Zinyowera, R.H. Moss, and D.J. Dokken, eds, *Climate Change 1995: Impacts, Adaptations and Mitigation of Climate Change: Scientific-Technical Analyses*, Cambridge University Press, Cambridge, UK, pp. 561–584.

McMichael, A.J., and Martens, W.J.M., 1995, Assessing health impacts of global environmental change: Grappling with scenarios, predictive models, and uncertainty, *Ecosystem Health*, **1**:15–25.

McMichael, A.J., Haines, A., Sloof, R., and Kovats, S., eds, 1996, *Climate Change and Human Health*, World Health Organisation, Geneva, Switzerland.

McNicoll, G., 1984, Consequences of rapid population growth, *Population and Development Review*, **10**(2):177–240.

McNicoll, G., 1990, Social organization and ecological stability under demographic stress, in G. McNicoll and M. Cain, eds, *Rural Development and Population: Institutions and Policy*, Oxford University Press, New York, NY, USA, pp. 147–160.

McNicoll, G., 1993, Malthusian Scenarios and Demographic Catastrophe, Population Council Research Division Working Paper No. 49, Population Council, New York, NY, USA.

McNicoll, G., 1995, On population growth and revisionism: Further questions, *Population and Development Review*, **21**(2):307–340.

McPhaden, M.J., 1999, The child prodigy of 1997-98, *Nature*, **398**:559–562.

Mearns, L.O., Katz, R.W., and Schneider, S.H., 1984, Extreme high-temperature events: Changes in their probabilities with changes in mean temperature, *Journal of Climatology and Applied Meteorology*, **23**(12):1601–1613.

Meehl, G.A., and Washington, W.M., 1993, South Asian summer monsoon variability in a model with doubled atmospheric carbon dioxide, *Science*, **260**:1101–1104.

Mehrota, D., and Mehrota, R., 1995, Climate change and hydrology with emphasis on the Indian subcontinent, *Hydrological Sciences*, **40**(2):231–242.

Mesle, F., 1993, The future of mortality, in R. Cliquet, ed., *The Future of Europe's Population: A Scenario Approach*, The Council of Europe, European Population Committee.

Meyerson, F.A.B., 1998, Population, carbon emissions, and global warming: The forgotten relationship at Kyoto, *Population and Development Review*, **24**(1):115–130.

M'Gonigle, R., 1999, Ecological economics and political ecology: Towards a necessary synthesis, *Ecological Economics*, **28**:11–26.

Mink, S., 1993, Poverty, Population, and Environment, *World Bank Discussion Paper 189*, The World Bank, Washington, DC, USA.

Minnis, P., Harrison, E.F., Stowe, L.L., Gibson, G.G., Denn, F.M., Doelling, D.R., Smith, W.L., 1993, Radiative climate forcing by the Mt. Pinatubo eruption, *Science*, **259**:1411–1415.

Mitchell, J.F.B., Johns, T.C., Gregory, J.M. and Tett, S.F.B., 1995, Climate response to increasing levels of greenhouse gases and sulphate aerosols, *Nature*, **376**:501–504.

Montgomery, M.R., 1996, Comments on men, women, and unintended pregnancy, in J.B. Casterline, R.D. Lee, and K.A. Foote, eds, *Fertility in the United States: New Patterns, New Themes, Supplement to Population and Development Review*, **22**: 100–106.

Montgomery, M., and Lloyd, C., 1996, Fertility and maternal and child health, in D. Ahlburg, A. Kelley, and K.O. Mason, eds, *The Impact of Population Growth on Well-Being in Developing Countries*, Springer-Verlag, Berlin, Germany, pp. 37–137.

Moomaw, W.R., and Tullis, D.M., 1998, Population, affluence, or technology: An empirical look at national carbon dioxide production, in B. Baudot and W. Moomaw, eds, *People and the Planet*, Macmillan, New York, NY, USA.

Morse, S.S., 1991, Emerging viruses: Defining the rules for viral traffic, *Perspectives in Biology and Medicine*, **34**(3):387–409.

Mortimore, M., 1989, Drought and Drought Response in the Sahel, Background paper prepared for the Committee on Human Consequences of Global Change, National Research Council, Washington, DC, USA.

Mortimore, M., 1993, Northern Nigeria: Land transformation under agricultural intensification, in C.J. Jolly and B.B. Torrey, eds, *Population and Land Use in Developing Countries*, National Academy Press, Washington, DC, USA, pp. 42–69.

Mount, T.D., 1994, Climate change and agriculture: A perspective on priorities for economic policy, *Climatic Change*, **27**:121–138.

Mueller, E., 1976, The economic value of children in peasant agriculture, in R. Ridker, ed., *Population and Development: The Search for Selective Interventions*, Johns Hopkins University Press, Baltimore, MD, USA.

Muhuri, P.K., and Menken, L., 1997, Adverse effects of next birth, gender, and family composition on child survival in rural Bangladesh, *Population Studies*, **51**:279–294.

Munich Re, 1997, *Annual Review of Natural Catastrophes*, Munich Reinsurances Corporation, Munich, Germany.

Murray, J., 1994, Nutrition, disease and health, in V. Ruttan, ed., *Health and Sustainable Agricultural Development: Perspectives on Growth and Constraints*, Westview Press, Boulder, CO, USA, pp. 65–70.

Murthy, W.S., Panda, M., and Parikh, J., 1997, Economic growth, energy demand, and carbon dioxide emissions in India: 1990–2020, *Environment and Development Economics*, **2**:173–193.

Myers, N., 1989, Population, environment, and conflict, in K. Davis and M.S. Bernstam, eds, *Resources, Environment and Population: Present Knowledge*, Oxford University Press, Oxford, UK, pp. 43–56.

Myers, N., 1993a, Tropical deforestation: The main deforestation fronts, *Environmental Conservation*, **20**(1):9–16.

Myers, N., 1993b, *Ultimate Security: The Environmental Basis of Political Stability*, W.W. Norton, New York, NY, USA.

Myers, N., 1994a, Population and the environment: The vital linkages, in *Population, Environment and Development: Proceedings of the United Nations Expert Group Meeting on Population, Environment and Development*, United Nations, New York, NY, USA, pp. 55–63.

Myers, N., 1994b, Environmental refugees in a globally warmed world, *BioScience*, **43**(11):752–761.

Naess, A., 1989, *Ecology, Community, and Lifestyle: Outline of an Ecosophy*, Cambridge University Press, Cambridge, UK.

Najafizadah, M., and Mennerick, L., 1988, Worldwide educational expansion from 1950 to 1980: The failure of the expansion of schooling in the developing countries, *Journal of Developing Areas*, **22**(3):333–358.

Nakićenović, N., Grübler, A., and McDonald, A., eds, 1998, *Global Energy Perspectives*, Cambridge University Press, Cambridge, UK.

NAS (National Academy of Sciences), 1991, *Policy Implications of Climate Change*, Washington, DC, USA.

Nash, L.L., and Gleick, P.H., 1991, Sensitivity of streamflow in the Colorado basin to climatic changes, *Journal of Hydrology*, **125**:221–241.

Ncftcl, A., Friedli, H., Moor, E., Lötscher, H., Oeschger, H., Siegenthaler, U., Stauffer, B., 1997, Historical carbon dioxide record from the Siple Station ice core, in *Trends Online: A Compendium of Data on Global Change*, Carbon Dioxide Information Analysis Center, Oak Ridge National Laboratory, Oak Ridge, TN, USA.

Nelson, R., 1988, Dryland management: The Desertification Problem, World Bank Environment Department Working Paper No. 8, The World Bank, Washington, DC, USA.

Neméc, J., 1990, Climate and development: Agricultural practices and water resources, in H.-J. Karpe, D. Otten, and S.C. Trinidale, eds, *Climate and Development: Climatic Change and Variability and the Resulting Social, Economic and Technological Implications*, Springer-Verlag, Berlin, Germany, pp. 397–408.

Neméc, J., 1994, Climate variability, hydrology and water resources: Do we communicate in the field?, *Hydrological Sciences*, **39**(3):193–197.

Nerem, R.S., 1999, Measuring very low frequency sea level variations using satellite altimeter data, *Global and Planetary Change*, **20**:157–171.

Nerlove, M., Razin, A., and Sadka, E., 1987a, *Household and Economy: Welfare Economics of Endogenous Fertility*, Academic Press, New York, NY, USA.

Nerlove, M., Razin, A., and Sadka, E., 1987b, Population policy and individual choice: A theoretical investigation, *Research Report 60*, International Food Policy Research Institute, Washington, DC, USA.

Newman, M.J., and Rood, R.T., 1977, Implications of solar evolution for the Earth's early atmosphere, *Science*, **198**:1035–1037.

Nicholls, N., Gruza, G.V., Jouzel, J., Karl, T.R., Ogallo, L.A., and Parker, D.E., 1996, Observed climate variability and change, in J.T. Houghton, L.G. Meira Filho, B.A. Callander, N. Harris, A. Kattenberg, and K. Maskell, eds, *Climate Change 1995: The Science of Climate Change*, Cambridge University Press, Cambridge, UK, pp. 133–192.

Nordhaus, W.D., 1993, Optimal greenhouse-gas reductions and tax policy in the "DICE" models, *American Economic Review, Papers and Proceedings*, **83**:313–317.

Nordhaus, W.D., 1994, *Managing the Global Commons*, MIT Press, Cambridge, MA, USA.

Nordhaus, W.D., 1996, A regional dynamic general-equilibrium model of alternative climate change strategies, *The American Economic Review*, **86**(4):741–765.

Nordhaus, W.D., and Boyer, J., 1998, What Are the External Costs of More Rapid Population Growth? Theoretical Issues and Empirical Estimates, Paper presented at the 150th Anniversary Meeting of the American Association for the Advancement of Science, Philadelphia, PA, USA, 15 February.

Nordhaus, W.D., and Yohe, G., 1983, Future Carbon Dioxide Emissions From Fossil Fuels, *Cowles Foundation Paper No. 580*, Yale University, New Haven, CT, USA.

Notestein, F.W., 1945, Population – The long view, in T.W. Schultz, ed., *Food for the World*, University of Chicago Press, Chicago, IL, USA, pp. 36–57.

OECD (Organisation for Economic Co-operation and Development), 1992, The economic costs of reducing CO_2 emissions, *OECD Economic Studies 19* (Winter).

OECD (Organisation for Economic Co-operation and Development), 1993, *Energy Statistics and Balances 1960–1991*, OECD, Paris, France.

OECD (Organisation for Economic Co-operation and Development), 1998, *Maintaining Prosperity in an Ageing Society*, OECD, Paris, France.

O'Hara, S.L., Street-Perrott, F., Alayne, F., and Burt, T.P., 1993, Accelerated soil erosion around a Mexican highland lake caused by pre-Hispanic agriculture, *Nature*, **362**(6415):48–50.

Oksanen, M., 1997, The moral value of biodiversity, *Ambio*, **26**(8):541–545.

Olshansky, S.J., Carnes, B.A., and Cassel, C., 1990, In search of Methuselah: Estimating the upper limits of human longevity, *Science*, **2**:634–640.

O'Neill, B.C., 1996, Greenhouse gases: Timescales, response functions, and the role of population growth in future emissions, PhD dissertation, Earth Systems Group, Department of Applied Science, New York University, New York, NY, USA.

O'Neill, B.C., The jury is still out on GWPs, *Climatic Change* (in press).

O'Neill, B.C., and Wexler, L., 1998, The greenhouse externality to childbearing: A sensitivity analysis, *Climatic Change* (in press).

O'Neill, B.C., Oppenheimer, M., and Gaffin, S., 1997, Measuring time in the greenhouse, *Climatic Change*, **37**:491–503.

Oppenheimer, M., 1998, Global warming and the stability of the West Antarctic Ice Sheet, *Nature*, **393**:325–332.

Orstom, E., 1990, *Governing the Commons: The Evolution of Institutions for Collective Action*, Cambridge University Press, Cambridge, UK.

Painuly, J.P., and Rev, S.M., 1998, Environmental dimensions of fertilizer and pesticide use: Relevance to Indian agriculture, *International Journal of Environment and Pollution*, **10**(2):273–288.

Panagoulia, D., 1992, Impacts of GISS-modelled climate changes on catchment hydrology, *Hydrological Sciences*, **37**(2):141–163.

Panayotou, T., 1993, Empirical tests and policy analysis of environmental degradation at different stages of economic development, *World Employment Programme Working Paper WEP 2-22/WP.238*, International Labour Organization, Geneva, Switzerland.

Panayotou, T., 1996, An inquiry into population, resources, and environment, in D. Ahlburg, A. Kelley, and K.O. Mason, eds, *The Impact of Population Growth on Well-Being in Developing Countries*, Springer-Verlag, Berlin, Germany, pp. 259–298.

Pandit, K., and Cassetti, E., 1989, The shifting patterns of sectoral labor allocation during development: Developed versus developing countries, *Annals of the Association of American Geographers*, **79**(3):329–344.

Parry, M.L., 1990, *Climate Change and World Agriculture*, Earthscan Publications, London, UK.

Parry, M.L., 1993, Climate change and the future of agriculture, *International Journal of Environment and Pollution*, **3**(1–3):13–30.

Parry, M.L., and Carter, T., 1990, Some strategies of response in agriculture to changes of climate, in H.-J. Karpe, D. Otten, and S.C. Trinidale, eds, *Climate and Development: Climatic Change and Variability and the Resulting Social, Economic and Technological Implications*, Springer-Verlag, Berlin, Germany, pp. 152–172.

Parry, M.L., and Rosenzweig, C., 1993, Food supply and the risk of hunger, *Lancet*, **342**:1345-1347.

Pearce, D.W., ed., 1991, *Blueprint 2: Greening the World Economy*, Earthscan, London, UK.

Pearce, D.W., Barbier, E., and Markandya, A., 1990, *Sustainable Development: Economics and Environment in the Third World*, Edward Elgar, London, UK.

Pearce, D.W., Cline, W.R., Achanta, A.N., Fankhauser, S., Pachauri, R.K., Tol, R.S.J., and Vellinga, P., 1996, The social costs of climate change: Greenhouse damage and the benefits of control, in J.P. Bruce, H. Lee, and E.F Haites, eds, *Climate Change 1995: Economic and Social Dimensions of Climate Change*, Cambridge University Press, Cambridge, UK, pp. 179–224.

Pepper, W., Leggett, J., Swart, J., Wasson, J., Edmonds, J., and Mintzer, I., 1992, Emissions Scenarios for the IPCC, Report prepared for the Intergovernmental Panel on Climate Change, Working Group I, Cambridge University Press, Cambridge, UK.

Percival, V., and Homer-Dixon, T.F., 1995, *Environmental Scarcity and Violent Conflict: The Case of South Africa*, American Association for the Advancement of Science, Washington, DC, USA.

Pernetta, J.C., 1992, Impacts of climate change and sea-level rise on small island states: National and international responses, *Global Environmental Change*, March:19–31.

Perry, C.J., 1995, Determinants of function and dysfunction in irrigation performance, and implications for performance improvement, *Water Resources Development*, 11(1):25–38.

Persson, A., and Munasinghe, M., 1995, Natural resource management and economy-wide policies in Costa Rica: A computable general equilibrium (CGE) modeling approach, *World Bank Economic Review*, 9(2):259–285.

Petit, J.R., Jouzel, J., Raynaud, D., Barkov, N.I., Barnola, J.M., Basile, I., Bender, M., Chappellaz, J., Davis, M., Delaygue, G., Delmotte, M., Kotlyakov, V.M., Legrand, M., Lipenkov, V.Y., Lorius, C., Pepin, L., Ritz, C., Saltzman, E., and Stievenard, M., 1999, Climate and atmospheric history of the past 420,000 years from the Vostok ice core, Antarctica, *Nature*, 399:429–436.

Pfaff, A., 1999, What drives deforestation in the Brazilian Amazon?, *Journal of Environmental Economics and Management*, 37:26–43.

Pingali, P., 1990, Institutional and environmental constraints to agricultural intensification, in G. McNicoll and M. Cain, eds, *Rural Development and Population: Institutions and Policy*, Oxford University Press, New York, NY, USA.

Pingali, P., and Binswanger, H., 1987, Population density and agricultural intensification: A study of the evolution of technologies in tropical agriculture, in P.G. Johnson and R.D. Lee, eds, *Population Growth and Economic Development: Issues and Evidence*, University of Wisconsin Press, Madison, WI, USA.

Pinstrup-Anderson, P., and Pandya-Lorch, R., 1998, Food security and sustainable use of natural resources: A 2020 vision, *Ecological Economics*, 26:1–10.

Pope, C.A., Bates, D.V., and Raizenne, M.E., 1995, Health effects of particulate air pollution: Time for reassessment, *Environmental Health Perspectives*, 103:472–480.

Portney, P.R., 1998, Applicability of cost-benefit analysis to climate change, in W.D. Nordhaus, ed., *Economics and Policy Issues in Climate Change*, Resources for the Future, Washington, DC, USA, pp. 111–127.

Portney, P.R., and Weyant, J.P., eds, 1999, *Discounting and Intergenerational Equity*, Resources for the Future, Washington, DC, USA.

Potrykowska, A., and Clarke, J.I., 1995, Population and environment in industrialized regions, *Geographica Polonica*, **64**, Institute of Geography and Spatial Organization, Polish Academy of Sciences, Warsaw, Poland.

Potter, P., 1991, Rapid population growth, the quality of health, and the quality of health care in developing countries, in *Consequences of Rapid Population Growth in Developing Countries: Proceedings of the United Nations/Institut national des études démographiques Expert Group Meeting*, 23–26 August 1988, Taylor and Francis, New York, NY, USA, pp. 219–241.

Prather, M., Derwent, R., Ehhalt, D., Fraser, P., Sanhueza, E., and Zhou, X., 1995, Other trace gases and atmospheric chemistry, in J.T. Houghton, L.G. Meira Filho, J. Bruce, H. Lee, B.A. Callander, E. Haites, N. Harris, and K. Maskell, eds, *Climate Change 1994: Radiative Forcing of Climate Change and An Evaluation of the IPCC IS92 Emission Scenarios*, Cambridge University Press, Cambridge, UK, pp. 73–126.

Preston, S.H., 1994, Population and environment: From Rio to Cairo, IUSSP Distinguished Lecture on Population and Development, International Union for the Scientific Study of Population, Liège, Belgium.

Pritchett, L.H., 1994a, Desired fertility and the impact of population policies, *Population and Development Review*, **20**(1):1–55.

Pritchett, L.H., 1994b, The impact of population policies: Reply, *Population and Development Review*, **20**(3):621–630.

Pritchett, L.H., 1997, Divergence, big time, *Journal of Economic Perspectives*, **11**(3):3–17.

Rabl, A., 1996, Discounting of long-term costs: What would future generations prefer us to do?, *Ecological Economics*, **17**:137–145.

Rahman, A., Robins, N., and Roncerel, A., eds, 1993, *Population Versus Consumption: Which is the Climate Bomb?*, Castle Cary Press, Somerset, UK.

Rahmstorf, S., 1997, Risk of sea-change in the Atlantic, *Nature*, **388**:825–826.

Rahmstorf, S., 1999, Shifting seas in the greenhouse?, *Nature*, **399**:523–524.

Rajagopalan, B., Lall, U., and Cane, M.A., 1997, Anomalous ENSO occurrences: An alternate view, *Journal of Climate*, **10**:2351–2357.

Ramanathan, V., Cicerone, R.J., Angh, H.B., and Kiehl, J.T., 1985, Trace gas trends and their potential role in climate change, *Journal of Geophysical Research*, **90**: 5547–5566.

Ramlogan, R., 1996, Environmental refugees: A review, *Environmental Conservation*, **23**(1):81–88.

Raskin, P.D., 1995, Methods for estimating the population contribution to environmental change, *Ecological Economics*, **15**:225–233.

Rayner, S., and Malone, E.L., eds, 1998, *Human Choice and Climate Change: An International Assessment* (4 vols.), Battelle Press, Columbus, OH, USA.

Reilly, J., 1994, Crops and climate change, *Nature*, **367**:118–119.

Reilly, J., 1996, Agriculture in a changing climate: Impacts and adaptations, in R.T. Watson, M.C. Zinyowera, R.H. Moss, and D.J. Dokken, eds, *Climate Change 1995: Impacts, Adaptations and Mitigation of Climate Change: Scientific-Technical Analyses*, Cambridge University Press, Cambridge, UK, pp. 427–467.

Renner, M., 1989, National Security: The Economic and Environmental Dimensions, Worldwatch Paper No. 89, Worldwatch Institute, Washington, DC, USA.

Revelle, R., and Suess, H.E., 1957, Carbon dioxide exchange between atmosphere and ocean and the question of an increase of atmospheric CO_2 during the past decades, *Tellus*, **IX**:18–27.

Richards, T., 1983, Weather, nutrition, and the economy: Short-run fluctuations in births, deaths, and marriages, France 1740–1909, *Demography*, **20**(2):197–212.

Rodgers, G.B., 1984, *Population and Poverty*, International Labour Organization, Geneva, Switzerland.

Rodgers, G.B., 1989, Introduction: Trends in urban poverty and labour market access, in G. Rodgers, ed., *Urban Poverty and the Labour Market*, International Labour Organization, Geneva, Switzerland.

Rogers, A., and Castro, L., 1981, *Model Migration Schedules*, RR-81-30, International Institute for Applied Systems Analysis, Laxenburg, Austria.

Rogers, D.T., and Packer, M.J., 1993, Vector-borne diseases, models and global change. *Lancet*, **342**:1282–1284.

Rogoff, M.H., and Rawlins, S.L., 1987, Food security: A technological alternative, *Bio-Science*, **37**:800–807.

Romer, P., 1990, Capital, labor and productivity, *Brookings Papers in Economic Activity: Microeconomics 1990*:337–367, Brookings Institution, Washington, DC, USA.

Romieu, I., Weitzenfeld, H., and Finkelman, J., 1990, Urban air pollution in Latin America and the Caribbean: Health perspectives, *World Health Statistics Quarterly*, **43**: 153–167.

Root, T.L., and Schneider, S.H., 1995, Ecology and climate: Research strategies and implications, *Science*, **269**:331–341.

Rosenberg, N.J., 1982, The increasing CO_2 concentration in the atmosphere and its implication for agricultural productivity, II, Effects through CO_2-induced climatic change, *Climatic Change*, **4**:239–254.

Rosenberg, N.J., 1992, Adaptation of agriculture to climate change, *Climatic Change*, **21**:385–405.

Rosenzweig, C., 1985, Potential CO_2-induced effects on North American wheat producing regions, *Climatic Change*, **7**:367–389.

Rosenzweig, C., and Hillel, D., 1998, *Climate Change and the Global Harvest: Potential Impacts of the Greenhouse Effect on Agriculture*, Oxford University Press, New York, NY, USA.

Rosenzweig, M., 1988, Human capital, population growth and economic data: Beyond correlations, *Journal of Policy Modelling*, **10**(1):83–111.

Rosenzweig, M., 1990, Population growth and human capital investments: Theory and evidence, *Journal of Political Economy*, **98**(5):S38–70.

Roseveare, D., Leibfritz, W., Fore, D., and Wurzel, E., 1996, Ageing populations, pension systems, and government budgets: Simulations for 20 OECD countries, *OECD Economics Department Working Papers No. 168*, OECD, Paris, France.

Rothman, D.S., and de Bruyn, S.M., 1998, Probing into the environmental Kuznets curve hypothesis, *Ecological Economics*, **25**:143–145.

Rudel, T., and Roper, J., 1996, Regional patterns and historical trends in tropical deforestation, 1976–1990: A qualitative comparative analysis, *Ambio*, **25**(3):160–166.

Ruttan, V.W., 1971, Technology and the environment, *American Journal of Agricultural Economics*, **53**:707–717.

Ruttan, V.W., 1991, Sustainable growth in agricultural production: Poetry, policy and science, in S. Vosti, T. Reardon, and W. von Urff, eds, *Agricultural Sustainability, Growth, and Poverty Alleviation: Issues and Policies*, International Food Policy Research Institute, Washington, DC, USA.

Ruttan, V.W., and Hayami, Y., 1991, Rapid population growth and institutional and technical change, in *Consequences of Rapid Population Growth in Developing Countries: Proceedings of the United Nations/Institut national d'études démographiques Expert Group Meeting*, New York, 23–26 August 1988, Taylor and Francis, New York, NY, USA, pp. 127–157.

Sagan, C., and Mullen, G.H., 1972, Earth and Mars: Evolution of atmospheres and surface temperatures, *Science*, **177**:52–56.

Sanderson, W., Packirisamy, K., and Wils, A., Multistate Population Model with HIV/AIDS and Education, Interim Report, International Institute for Applied Systems Analysis, Laxenburg, Austria (forthcoming).

Santer, B.D., Wigley, T.M.L., Barnett, T.P., and Aryamba, E., 1996, Detection of climate change and attribution of causes, in J.T. Houghton, L.G. Meira Filho, B.A. Callander, N. Harris, A. Kattenberg, and K. Maskell, eds, *Climate Change 1995: The Science of Climate Change*, Cambridge University Press, Cambridge, UK, pp. 407–444.

Scheiber, J., and Shoven, J., 1994, The Consequences of Population Aging on Private Pension Fund Saving and Asset Markets, NBER Working Paper No. 466, National Bureau for Economic Research, Cambridge, MA, USA.

Schelling, T.C., 1995, Intergenerational discounting, *Energy Policy*, **23**(4/5):395–401.

Schimel, D., Alves, D., Enting, I., Heimann, M., Joos, F., Raynaud, D., Wigley, T., Prather, M., Derwent, R., Ehhalt, D., Fraser, P. Sanhueza, E., Zhou, X., Jonas, P., Charlson, R., Rodhe, H., Sadasivan, S., Shine, K.P., Fouquart, Y., Ramaswamy, V., Solomon, S., Srinivasan, J., Albritton, D., Isaksen, I., Lal, M., and Wuebbles, D., 1996, Radiative forcing of climate change, in J.T. Houghton, L.G. Meira Filho, B.A. Callander, N. Harris, A. Kattenberg, and K. Maskell, eds, *Climate Change 1995: The Science of Climate Change*, Cambridge University Press, Cambridge, UK, pp. 65–132.

Schneider, S.H., 1997, Integrated assessment modeling of global climate change: Transparent rational tool for policy making or opaque screen hiding value-laden assumptions?, *Environmental Modeling and Assessment*, **2**(4):229–248.

Schultz, T.P., 1987, School expenditures and enrollments, 1960–1980: The effects of income, prices and population growth, in D.G. Johnson and R. Lee, eds, *Population Growth and Economic Development: Issues and Evidence*, University of Wisconsin Press, Madison, WI, USA, pp. 413–476.

Scott, M.J., 1996, Human settlements in a changing climate: Impacts and adaptation, in R.T. Watson, M.C. Zinyowera, R.H. Moss, and D.J. Dokken, eds, *Climate Change 1995: Impacts, Adaptations and Mitigation of Climate Change: Scientific-Technical Analyses*, Cambridge University Press, Cambridge, UK, pp. 399–426.

Sen, A., 1981, *Poverty and Famines: An Essay on Entitlement and Deprivation*, Clarendon Press, Oxford, UK.

Sen, G., 1994a, Women, poverty and population: Issues for the concerned environmentalist, in L. Arizpe, M.P. Stone, and D.C. Major, eds, *Population and the Environment: Rethinking the Debate*, Westview Press, Boulder, CO, USA, pp. 67–86.

Sen, G., 1994b, Women's empowerment and human rights: The challenge to policy, in *Population: The Complex Reality*, The Royal Society, University Press, Cambridge, UK.

Sen, G., 1995, The World Programme of Action: A new paradigm for population policy, *Environment*, **37**(1):10–15, 34–37.

Serageldin, I., 1995, Water resources management: A new policy for a sustainable future, *Water Resources Development*, **11**(3):221–222.

Severinghaus, J.P., Sowers, T., Brook, E.J., Alley, R.B., Bender, M.L., 1998, Timing of abrupt climate change at the end of the Younger Dryas interval from thermally fractionated gases in polar ice, *Nature*, **391**:141–146.

Shafik, N., and Bandyopadhyay, S., 1992, Economic Growth and Environmental Quality: Time Series and Cross-country Evidence, World Bank Policy Research Paper WPS 904, The World Bank, Washington, DC, USA.

Shaw, R.P., 1989, Rapid population growth and environmental degradation: Ultimate versus proximate factors, *Environmental Conservation*, **16**(3):199–208.

Shaw, R.P., 1993, Review of Harrison 1992, *Population and Development Review*, **19**(1):189–192.

Sheffield, J., 1998, World population growth and the role of annual energy use per capita, *Technological Forecasting and Social Change*, **59**:55–87.

Shindell, D., Rind, E. Balachandran, N., Lean, J. and Lonergan, P., 1999a, Solar cycle variability, ozone, and climate, *Science*, **284**:305–308.

Shindell, D.T., Miller, R.L., Schmidt, G.A., and Pandolfo, L., 1999b, Simulation of recent northern winter climate trends by greenhouse gas forcing, *Nature*, **399**:452–455.

Shine, K.P., Fouquart, Y., Ramaswamy, V., Solomon, S., Srinivasan, J., 1995, Radiative forcing, in J.T. Houghton, L.G. Meira Filho, B.A. Callander, N. Harris, A. Kattenberg, and K. Maskell, eds, *Climate Change 1994: Radiative Forcing of Climate Change and An Evaluation of the IPCC IS92 Emission Scenarios*, Cambridge University Press, Cambridge, UK, pp. 163–204.

Shine, K.P., and Forster, M. de F., 1999, The effect of human activity on radiative forcing of climate change: A review of recent developments, *Global and Planetary Change*, **20**:205–225.

Shope, R., 1991, Global climate change and infectious diseases, *Environmental Health Perspectives*, **96**:171–174.

Shuter, B.J., and Post, J.R., 1990, Climate change, population viability, and the zoogeography of temperate fishes, *Transactions of the American Fisheries Society*, **1/9**: 314–336.

Simon, J., 1976, Population growth may be good for LDCs in the long run: A richer simulation model, *Economic Development and Cultural Change*, **24**:309–337.

Simon, J., 1981, *The Ultimate Resource*, Princeton University Press, Princeton, NJ, USA.

Simon, J., 1989, *The Economic Consequences of Immigration*, Basil Blackwell, Boston, MA, USA.

Sinha, S.K., and Swaminathan, M.S., 1991, Reforestation, climate change, and sustainable nutrition security: A case study of India, *Climatic Change*, **19**:201–209.

Sivakumar, M.V.K., 1992, Climate change and implications for agriculture in Niger, *Climatic Change*, **20**:297–312.

Skea, J., 1999, Flexibility, emissions trading and the Kyoto Protocol, in *Pollution for Sale: Emissions Trading and Joint Implementation*, S. Sorrell and J. Skea, eds, Edward Elgar, Cheltenham, UK, pp. 354–379.

Smil, V., 1990, Planetary warming: Realities and responses, *Population and Development Review*, **16**(1):1–29.

Smil, V., 1991, Population growth and nitrogen: An exploration of a critical existential link, *Population and Development Review*, **17**(4):569–601.

Smil, V., 1993, *China's Environmental Crisis: An Inquiry into the Limits of National Development*, M.E. Sharpe, New York, NY, USA.

Smil, V., 1999, Nitrogen in crop production: An account of global flows, *Global Biogeochemical Cycles*, **13**:647–662.

Smit, B., and Yunlong, C., 1996, Climate change and agriculture in China, *Global Environmental Change*, **6**(3):205–214.

Smith, D.R., 1994, Change and variability in climate and ecosystem decline in Aral Sea basin deltas, *Post-Soviet Geography*, **35**(3):142–165.

Smith, D.R., 1995, Environmental security and shared water resources in post-Soviet Central Asia, *Post Soviet Geography*, **36**(6):351–370.

Smith, I.N., Dix, M., and Allan, R.J., 1997, The effect of greenhouse SSTs on ENSO simulations with an AGCM, *Journal of Climate*, **10**:342–352.

Smith, S., and Wigley, T.M.L., Global Warming Potentials: 1. Climatic implications of emissions reductions, *Climatic Change* (in press).

Solley, W.B., Merk, C.F., and Pierce, R.R., 1989, Estimated Use of Water in the United States in 1985, United States Geological Survey Circular No. 1004, Government Printing Office, Washington, DC, USA.

Standing, G., 1978, *Labour Force Participation and Development*, International Labour Organization, Geneva, Switzerland.

Stanley, D.J., and Warne, A.G., 1993, Nile Delta: Recent geological evolution and human impact, *Science*, **260**(30 April):628–634.

Starr, J.R., 1991, Water wars, *Foreign Policy*, **70**(2):17–36.

Steig, E.J., Brook, E.J., White, J.W.C., Sucher, C.M., Bender, M.L., Lehman, S.J., Morse, D.L., Waddington, E.D., and Clow, G.D., 1998, Synchronous climate changes in Antarctica and the North Atlantic, *Science*, **282**:92–95.

Steinmann, G., and Komlos, J., 1988, Population growth and economic development in the very long run: A simulation model of three revolutions, *Mathematical Social Sciences*, **16**(1):44–63.

Stern, D.I., 1998, Progress on the environmental Kuznets curve?, *Environment and Development Economics*, **3**:173–196.

Stern, P., Young, O., and Druckman, D., eds, 1992, *Global Change: Understanding the Human Dimensions*, National Academy Press, Washington, DC, USA.

Stevenson, D.S., Johnson, C.E., Collins, W.J., Derwent, R.G., Shine, K.P., and Edwards, J.M., 1998, Evolution of tropospheric ozone radiative forcing, *Geophysical Research Letters*, **25**:3819–3822.

Stocker, T.F., and Schmittner, A., 1997, Influence of CO_2 emission rates on the stability of the thermohaline circulation, *Nature*, **388**:862–865.

Strauss, J., and Thomas, D., 1998, Health, nutrition and economic development, *Journal of Economic Literature*, **36**:766–817.

Stripple, J., 1998, Securitizing the Risks of Climate Change: International Innovations in the Insurance of Catastrophic Risk, IR-98-98, International Institute for Applied Systems Analysis, Laxenburg, Austria.

Strzepek, K., Niemann, J., Somlyódy, L., and Kulshreshtra, S., 1995, A Global Assessment of National Water Resources Vulnerabilities: Sensitivities, Assumptions, and Driving Forces, unpublished manuscript.

Suhrke, A., 1993, Pressure points: Environmental degradation, migration and conflict, *Occasional Paper Series on Environmental Change and Acute Conflict*, Peace and Conflict Studies Program, University of Toronto and International Security Studies Program, American Academy of Arts and Sciences, Toronto, Canada, and Cambridge, MA, USA.

Sutherst, R.W., 1991, Pest risk analysis and the greenhouse effect, *Review of Agricultural Entomology*, **79**(11/12):1177–1187.

Tabah, L., 1989, From one demographic transition to another, *Population Bulletin of the United Nations*, **28**:1–24.

Teitelbaum, M., and Winter, J., 1985, *The Fear of Population Decline*, Academic Press, Orlando, FL, USA.

Tett, S., 1995, Simulation of El Niño-Southern Oscillation-like variability in a global AOGCM and its response to CO_2 increase, *Journal of Climate*, **8**:1473–1502.

Thomas, N., 1991, Land, fertility, and the population establishment, *Population Studies*, **45**:379–397.

Thomas, N., 1992, Review of UNFPA, population, resources, and the environment: The critical challenges, *Population Studies*, **46**(3):559–560.

Thompson, J.R., and Hollis, G.E., 1995, Hydrological modelling and the sustainable development of the Hadejia-Nguru Wetlands, Nigeria, *Hydrological Sciences*, **40**(1):97–116.

Thurman, E.A., Goolsby, D.A., Meyer, M.T., Mills, M.S., Pomes, M.L., and Kolpin, D.W., 1991, Herbicides in surface waters of the United States: The effect of spring flush, *Environmental Science and Technology*, **25**:1794–1796.

Thurman, E.A., Goolsby, D.A., Meyer, M.T., Mills, M.S., Pomes, M.L., and Kolpin, D.W., 1992, A reconnaissance study of herbicides and their metabolites in surface water of the mid-western United States using immunoassay and gas chromatography/mass spectrometry, *Environmental Science and Technology*, **26**:2440–2447.

Tiffen, M., and Mortimore, M., 1992, Environment, population growth and productivity in Kenya, *Development Policy Review*, **10**(4):359–387.

Tiffen, M., Mortimore, M., and Gichuki, F.N., 1994, *More People, Less Erosion: Environmental Recovery in Kenya*, John Wiley, New York, NY, USA.

Timmerman, A., Oberhuber, J., Bacher, A., Esch, M., Latif, M., and Roeckner, E., 1999, Increased El Niño frequency in a climate model forced by future greenhouse warming, *Nature*, **398**:694–697.

Trenberth, K.E., 1997a, The definition of El Niño, *Bulletin of the American Meteorological Society*, **78**:2771–2777.

Trenberth, K.E., 1997b, The use and abuse of climate models, *Nature*, **386**:131–133.

Trenberth, K.E., and Hoar, T.J., 1996, The 1990-95 El Niño–Southern Oscillation event: Longest on record, *Geophysical Research Letters*, **23**:57–60.

Trenberth, K.E., and Hoar, T.J., 1997, El Niño and climate change, *Geophysical Research Letters*, **24**:3057–3060.

Trexler, M.C., and Kosloff, L.H., 1998, The 1997 Kyoto Protocol: What does it mean for project-based climate change mitigation?, *Mitigation and Adaptation Strategies for Global Change*, **3**:1–58.

Turner, D., Giorno, C., de Serres, A., Vourc'h, A., and Richardson, P., 1998, The macro-economic implications of ageing in a global context, *OECD Ageing Working Paper AWP 1.2*, OECD, Paris, France.

UNFPA (United Nations Population Fund), 1991, *Population, Resources, and the Environment: The Challenges Ahead*, United Nations, New York, NY, USA.

United Nations, 1973, Declaration of the UN Conference on the Human Environment, 16 June 1972, UN Doc A/CONf.48/14/Rev.1.

United Nations, 1987, *Fertility Behavior in the Context of Development*, Population Studies 100.ST/ESA/SER.A/100, Department of International Economic and Social Affairs, New York, NY, USA.

United Nations, 1993, *World Population Prospects: The 1992 Revision*, United Nations, New York, NY, USA.

United Nations, 1996, Document: FCCC/CP/1996/12, Second compilation and synthesis of first national communications from Annex I parties.

United Nations, 1997, *World Population Prospects: The 1996 Revision*, United Nations, New York, NY, USA.

United Nations, 1998a, *World Urbanization Prospects: The 1996 Revision, Estimates and Projections of Urban and Rural Populations and of Urban Agglomerations*, ST/ESA/SER.A/170, Department of Economic and Social Affairs, Population Division, United Nations, New York, NY, USA.

United Nations, 1998b, Review of the Implementation of Commitments and of other Provisions of the Convention: National Communications from Parties Included in Annex I to the Convention: Second Compilation and Synthesis of Second National Communications, FCCC/CP/1998/11, United Nations, New York, NY, USA.

United Nations Secretariat, 1989, Correlates of fertility in selected developing countries, *Population Bulletin of the United Nations*, **28**:95–106.

United Nations Secretariat, 1991, Relationships between population and environment in rural areas of developing countries, *Population Bulletin of the United Nations*, **31/32**:52–69.

United Nations Secretariat, 1994, Population and environment: An overview, in *Population, Environment and Development: Proceedings of the United Nations Expert Group Meeting on Population, Environment and Development*, United Nations, New York, NY, USA, pp. 23–38.

USDA (United States Department of Agriculture), 1993, *Expenditures on a Child by Families*, Family Economics Research Group, Hyattsville, MD, USA.

US National Research Council, 1984, *Causes and Effects of Changes in Stratospheric Ozone: Update 1983*, National Academy Press Washington, DC, UK.

US National Research Council, 1986, *Population Growth and Economic Development: Policy Questions*, National Academy Press, Washington, DC, USA.

Valkonen, T., 1991, Assumptions about mortality trends in industrialized countries: A survey, in W. Lutz, ed., *Future Demographic Trends in Europe and North America: What Can We Assume Today?*, Academic Press, London, UK, pp. 3–25.

van de Kaa, D.J., 1993, European migration at the end of history, *European Review*, **1**(1):87–108.

van de Walle, E., 1992, Fertility transition, conscious choice and numeracy, *Demography*, **29**(4):487–502.

van Imroff, E., 1988, Age structure, education, and the transmission of technical change, *Journal of Population Economics*, **1**(3):167–181.

Vaupel, J.W., and Lundström, H., 1996, The future mortality at older ages in developed countries, in W. Lutz, ed., *The Future Population of the World: What Can We Assume Today?*, Earthscan, London, UK, pp. 278–295.

Vaupel, J.W., Carey, J.R., Christensen, K., Johnson, T.E., Yashin, A.I., Holm, N.V., Iachine, I.A., Kannisto, V., Khazaeli, A.A. Liedo, P., Longo, V.D., Zeng, Y., Manton, K.G., and Curtsinger, J.W., 1998, Biodemographic trajectories of longevity, *Science*, **280**:855–860.

Vincent, J., 1992, The tropical timber trade and sustainable development, *Science*, **256** (19 June):1651–1655.

Vishnevsky, A., 1991, Demographic revolution and the future of fertility: A systems approach, in W. Lutz, ed., *Future Demographic Trends in Europe and North America: What Can We Assume Today?*, Academic Press, London, UK, pp. 257–270.

Vringer, K., and Blok, K., 1995, The direct and indirect energy-requirements of households in the Netherlands, *Energy Policy*, **23**(10):893–902.

Waggonner, P.E., 1994, How much land can ten billion people spare for nature?, *Task Force Report*, No. 121, Council for Agricultural Science and Technology, Ames, IA, USA.

Walker, J.C.G., Hays, P.B., and Kasting, J.F., 1981, A negative feedback mechanism for the long-term stabilization of Earth's surface temperature, *Journal of Geophysical Research*, **86**:9776–9782.

Walker, R.T., 1987, Land use transition and deforestation in developing countries, *Geographical Analysis*, **19**(1):18–30.

Wang, X.H., and Feng, Z.M., 1997, Rural household energy consumption in Yangzhong County of Jiangsu Province in China, *Energy*, **22**(12):1159–1162.

Warrick, R.A., 1980, Drought in the Great Plains: A case study of research on climate and society in the USA, in J. Ausubel and A. Biswas, eds, *Climatic Constraints and Human Activities*, Pergamon Press, New York, NY, USA, pp. 92–123.

Warrick, R.A., and Rahman, A., 1992, Future sea level rise: Environmental and socio-political considerations, in I. Mintzer, ed., *Confronting Climate Change: Risks, Implications, and Responses*, Cambridge University Press, Cambridge, UK, pp. 97–112.

Warrick, R.A., LeProvost, C., Meier, M.F., Oerlemans, J., and Woodworth, P.L., 1996, Changes in sea level, in J.T. Houghton, L.G. Meira Filho, B.A. Callander, N. Harris, A. Kattenberg, and K. Maskell, eds, *Climate Change 1995: The Science of Climate Change*, Cambridge University Press, Cambridge, UK, pp. 359–406.

Watson, R.T., Zinyowera, M.C., and Moss, R.H., eds, 1996, *Climate Change 1995: Impacts, Adaptations and Mitigation of Climate Change: Scientific-Technical Analyses*, Cambridge University Press, Cambridge, UK.

WCED (World Commission on Environment and Development), 1987, *Our Common Future*, Oxford University Press, Oxford, UK.

Weart, S.R., 1997, The discovery of the risk of global warming, *Physics Today*, January: 34–40.

Weinberger, M.B., 1994. Recent trends in contraceptive use, *Population Bulletin of the United Nations*, **36**:55–80.

Weiner, J., 1990, *The Next One Hundred Years: Shaping the Fate of Our Living Earth*, Bantam Books, New York, NY, USA.

Weir, D., 1991, A historical perspective on the economic consequences of rapid population growth, in *Consequences of Rapid Population Growth in Developing Countries: Proceedings of the United Nations/Institut national d'études démographiques Expert Group Meeting*, New York, 23–26 August 1988, Taylor and Francis, New York, NY, USA, pp. 41–66.

Wentz, F.J., and Schabel, M., 1998, Effects of orbital decay on satellite-derived lower-tropospheric temperature trends, *Nature*, **394**:661–664.

Wescoat, J.L., Jr., 1991, Managing the Indus River basin in light of climate change: Four conceptual approaches, *Global Environmental Change*, December:381–395.

Westing, A.H., ed., 1986, *Global Resources and International Conflict: Environmental Factors in Strategic Policy and Action*, Oxford University Press, New York, NY, USA.

Westing, A.H., 1992, Environmental refugees: A growing category of displaced persons, *Environmental Conservation*, *19*:201–207.

Westing, A.H., 1994, Population, desertification, and migration, *Environmental Conservation*, **21**(2):110–115.

Westoff, C., 1991, The return to replacement fertility: A magnetic force?, in W. Lutz, ed., *Future Demographic Trends in Europe and North America: What Can We Assume Today?*, Academic Press, London, UK, pp. 227–233.

Westoff, C., 1995, International population policy, *Society*, May/June:11–15.

Westoff, C., 1996, Reproductive preferences and future fertility in developing countries, in W. Lutz, ed., *The Future Population of the World: What Can We Assume Today?*, Earthscan, London, UK, pp. 73–87.

Wexler, L., 1996a, Decomposing Models of Demographic Impact on the Environment, WP-96-85, International Institute for Applied Systems Analysis, Laxenburg, Austria.

Wexler, L., 1996b, Improving population assumptions in greenhouse gas emissions models, WP-96-99, International Institute for Applied Systems Analysis, Laxenburg, Austria.

Wexler, L., 1996c, The Greenhouse Externality to Childbearing, unpublished manuscript.

Weyant, J., 1996, Integrated assessment of climate change: An overview and comparison of approaches and results, in J.P. Bruce, H. Lee, and E.F. Haites, eds, *Climate Change 1995: Economic and Social Dimensions of Climate Change*, Cambridge University Press, Cambridge, UK, pp. 367–396.

Whelpton, P.K., 1936, An empirical method of calculating future population, *Journal of the American Statistical Association*, **31**:457–473.

Wigley, T.M.L., 1995, Global-mean temperature and sea level consequences of greenhouse gas concentration stabilization, *Geophysical Research Letters*, **22**(1):45–48.

Wigley, T.M.L., 1999, *The Science of Climate Change: Global U.S. Perspectives*, Pew Center on Global Climate Change, Washington, DC, USA.

Wigley, T.M.L., Richels, R., and Edmonds, J.A., 1995, Economic and environmental choices in the stabilization of CO_2 concentrations: Choosing the "right" emissions pathway, *Nature*, **379**:240–243.

Wigley, T.M.L., Jain, A.K., Joos, F., Nyenzi, B.S., and Shukla, P.R., 1997, in J.T. Houghton, L.G. Meira Filho, D.J. Griggs, and M. Noguer, eds, *IPCC Technical Paper 4: Implications of Proposed CO_2 Emissions Limitations*, Intergovernmental Panel on Climate Change, Cambridge University Press, Cambridge, UK.

Wilkinson, R.G., 1994, The epidemiological transition: From material scarcity to social disadvantage?, *Dædalus*, **123**(4):61–77.

Winters, R., Murgai, R., Sadoulet, E., de Jamry, A., and Frisvold, G., 1998, Economic and welfare impacts of climate change on developing countries, *Environmental and Resource Economics*, **12**:1–24.

Wolf, A.P., and Ying-Chang, C., 1994, Fertility and women's labor: Two negative (but instructive) findings, *Population Studies*, **48**(3):427–433.

Wolf, J., and Van Diepen, C.A., 1995, Effects of climate change on grain yield potential in the European Community, *Climatic Change*, **29**:299–331.

Wolman, G., 1993, Population, land use, and environment: A long history, in C. Jolly and B. Torrey, eds, *Population Growth and Land Use Change in Developing Countries*, National Academy Press, Washington, DC, USA, pp. 15–29.

Wood, R.A., Keen, A.B., Mitchell, J.F.B., and Gregory, J.M., 1999, Changing spatial structure of the thermohaline circulation in response to atmospheric CO_2 forcing in a climate model, *Nature*, **399**:572–575.

World Bank, 1992, *World Development Report 1992: Development and Environment*, The World Bank, Washington, DC, USA.

World Bank, 1993, *World Development Report 1993: Investing in Health*, The World Bank, Washington, DC, USA.

World Bank, 1994, *Averting the Old Age Crisis*, The World Bank, Washington, DC, USA.

Wuebbles, D.J., Jain, A.K., Patten, K.O., and Grant, K.E., 1995, Sensitivity of direct global warming potentials to key uncertainties, *Climatic Change*, **29**:265–297.

Wynne, B., 1992, Uncertainty and environmental learning: Reconceiving science and policy in the preventive paradigm, *Global Environmental Change – Human and Policy Dimensions*, **2**(2):111–127.

Yang, C., and Schneider, S.H., 1998, Global carbon dioxide emission scenarios: Sensitivity to social and technological factors in three regions, *Mitigation and Adaptation Strategies for Global Change*, **2**:373–404.

Ye, X., and Taylor, J.E., 1995, The impact of income growth on farm household nutrient intake: A case study of a prosperous rural area in Northern China, *Economic Development and Cultural Change*, **43**(4):805–819.

Yousif, H.M., 1995, Population, biomass and the environment in central Sudan, *International Journal of Sustainable Development and World Ecology*, **2**:54–69.

Zlotnik, H., 1996, Migration to and from developing regions: A review of past trends, in W. Lutz, ed., *The Future Population of the World: What Can We Assume Today?*, Earthscan, London, UK, pp. 299–335.

Index